Springer Theses

Recognizing Outstanding Ph.D. Research

Aims and Scope

The series "Springer Theses" brings together a selection of the very best Ph.D. theses from around the world and across the physical sciences. Nominated and endorsed by two recognized specialists, each published volume has been selected for its scientific excellence and the high impact of its contents for the pertinent field of research. For greater accessibility to non-specialists, the published versions include an extended introduction, as well as a foreword by the student's supervisor explaining the special relevance of the work for the field. As a whole, the series will provide a valuable resource both for newcomers to the research fields described, and for other scientists seeking detailed background information on special questions. Finally, it provides an accredited documentation of the valuable contributions made by today's younger generation of scientists.

Theses are accepted into the series by invited nomination only and must fulfill all of the following criteria

- They must be written in good English.
- The topic should fall within the confines of Chemistry, Physics, Earth Sciences, Engineering and related interdisciplinary fields such as Materials, Nanoscience, Chemical Engineering, Complex Systems and Biophysics.
- The work reported in the thesis must represent a significant scientific advance.
- If the thesis includes previously published material, permission to reproduce this must be gained from the respective copyright holder.
- They must have been examined and passed during the 12 months prior to nomination.
- Each thesis should include a foreword by the supervisor outlining the significance of its content.
- The theses should have a clearly defined structure including an introduction accessible to scientists not expert in that particular field.

More information about this series at http://www.springer.com/series/8790

Jelena Aleksić

Optimized Dark Matter Searches in Deep Observations of Segue 1 with MAGIC

Doctoral Thesis accepted by
Universitat Autònoma de Barcelona, Barcelona, Spain

Springer

Author
Dr. Jelena Aleksić
Experimental Division
Institut de Física d'Altes Energies (IFAE)
Edifici Cn. Universitat Autònoma de Barcelona
Bellaterra (Barcelona)
Spain

Supervisors
Dr. Javier Rico
Institut de Física d'Altes Energies
Edifici Cn. Universitat Autònoma de Barcelona
Bellaterra (Barcelona)
Spain

Dr. Manel Martinez
Institut de Física d'Altes Energies
Edifici Cn. Universitat Autònoma de Barcelona
Bellaterra (Barcelona)
Spain

ISSN 2190-5053 ISSN 2190-5061 (electronic)
Springer Theses
ISBN 978-3-319-23122-8 ISBN 978-3-319-23123-5 (eBook)
DOI 10.1007/978-3-319-23123-5

Library of Congress Control Number: 2015951766

Springer Cham Heidelberg New York Dordrecht London

Printed on acid-free paper

Springer International Publishing AG Switzerland is part of Springer Science+Business Media (www.springer.com)

Mojim roditeljima

Supervisors' Foreword

The existence of dark matter in the Universe is supported by a vast body of experimental evidence, mainly involving its gravitational effects on the ordinary (baryonic) matter. For instance, the dynamics of cosmic structures, like galaxies or galaxy clusters, cannot be explained unless a dominant dark matter component is invoked. In addition, without dark matter, the Universe would not contain stars—or life as we know it—since gravitational attraction from the baryonic components alone could not have counteracted the effects of the cosmic expansion, and therefore would not have collapsed to form galaxies. Dark matter also left its imprint in the spectrum of the oldest light produced in the Universe—380,000 years after the Big Bang—that today we observe as the cosmic microwave background. Still, after many years of theoretical and experimental endeavors, we continue to seek the answer to the mystery of the nature of dark matter.

The scientific relevance of the detection and identification of dark matter cannot be overestimated. It would be a discovery of the utmost importance, with profound implications on our basic views of nature and, in particular, on fundamental questions regarding particle physics, astrophysics, and cosmology. To briefly mention a few: the identification of the dark matter particle would be an unassailable evidence for physics beyond the Standard Model, leading to a major revolution in the field; measurements of dark matter distribution in galaxies and galaxy clusters would improve our understanding of formation, evolution and dynamics of large structures, thus shedding light on the history of the Universe; dark matter is also the earliest cosmic relic we could detect and, as such, the knowledge of its properties is the key for our comprehension of the early epochs of the Universe prior to nucleosynthesis. Within the present theoretical and experimental landscapes, there are realistic chances for a fundamental discovery of the greatest scientific importance, as well as for constraining the different theoretical scenarios in the most stringent way.

This work represents a leap forward in dark matter searches with Cherenkov telescopes—instruments that look for gamma rays expected from annihilation or decay of dark matter particles in the local Universe. Dr. Aleksić introduces a new analysis technique that optimizes, for the first time in the field, the sensitivity of these detectors for signals of dark matter origin. This approach has become the standard in the Cherenkov astronomy, and today it is being used in a great variety of searches involving different promising dark matter targets. In addition, using the MAGIC telescopes at La Palma, Dr. Aleksić performed the deepest observations of a dwarf satellite galaxy by any Cherenkov telescope till date. By analyzing these extensive data with the full likelihood method, Dr. Aleksić has produced the most constraining limits to the properties of dark matter particles obtained from dwarf galaxy observations by any ground-based gamma-ray instrument, making her results a reference in the field.

It has been a real pleasure and a privilege for me to collaborate with Dr. Aleksić and witness the development of this extensive, comprehensive, and influential work. I hope that your experience from reading this book will be equally beneficial.

Barcelona
May 2015

Dr. Javier Rico
Dr. Manel Martinez

Preface

Dark matter is believed to be the dominant form of matter in the Universe, representing almost 85 % of the total mass density. Its existence is supported by abundant observational evidence on all scales, while independently motivated by the theoretical extensions to the Standard Model. Yet, despite the intensive efforts over the last decades, the nature of dark matter is still unknown, posing as one of the most exciting questions of modern science.

Current searches for dark matter constituents are primarily directed towards some new and exotic, weakly interacting massive particles, of non-baryonic nature that were produced thermally in the early Universe. The principal detection strategies of the existing experiments are based on (viable) possibilities that the dark matter particles could be 'seen' through some relation with the Standard Model ones: either by being created in collisions of hadrons or leptons (production in colliders), or by scattering off nuclei (direct detection), or by annihilating and decaying into Standard Model particles (indirect detection).

This work addresses the last approach, by searching for photons of dark matter origin. Considering the expectations for the dark matter particle mass (tens of GeV to few TeV), the resulting photons may be very energetic and in the gamma ray domain, a range currently best explored by the Large Telescope Array on board the Fermi satellite and by the ground-based Imaging Air Cherenkov Telescopes. Gamma rays are particularly interesting as dark matter messengers: not only do they travel undeflected through the Universe, thus directly pointing back to the place of their creation, but they also carry information about the properties of the dark matter particle itself. Namely, the photon spectrum from dark matter annihilation or decay could have some characteristic features, that cannot be imitated by astrophysical sources, and whose detection would unambiguously confirm the dark matter existence. Furthermore, these gamma rays are also encoded with details about the dark matter particle, like its mass, interactions cross section, and lifetime.

Here are presented results from indirect searches in a highly dark matter-dominated dwarf spheroidal galaxy Segue 1 with the MAGIC (Major Atmospheric Gamma ray Imaging Cherenkov) telescopes. The observations, carried out during the course of three years, have resulted in 158 h of good-quality data, making this the deepest survey of any dwarf spheroidal by any Cherenkov Telescope so far. Data are analyzed with the *full likelihood method*, a dedicated approach developed as a part of this work, which improves the sensitivity of the search for spectral features expected in dark matter signal. The results of the analysis are then used to infer constraints on the particle physics properties of dark matter, assuming different annihilation and decay final states.

The outline of the thesis can be summarized as follows:

- Chapter 1 introduces the dark matter paradigm: what are the astrophysical and cosmological evidence supporting the existence of dark matter and how can they be reconciled with our current image of the Universe. This Chapter also brings a short review of some of the best motivated candidates for dark matter particle, with accent on those that are of particular interest for this work.
- Chapter 2 is devoted to dark matter searches. It begins with the presentation of different strategies currently employed by various experiments and their most worth noting results, to continue with the more detailed description of indirect searches. Special attention is given to the highly energetic photons as search messengers: what signal should be expected, where to look for it and with which instruments.
- Chapter 3 introduces the tool used in this work for dark matter searches—the MAGIC telescopes. Chapter is divided into two parts: one, describing the technical properties of the system and the other, characterizing its standard analysis chain.
- Chapter 4 presents one of the original scientific contributions of this work—the development of the full likelihood approach, an analysis method optimized for recognition of spectral features expected from photons of dark matter origin. After the introduction and characterization of the method, its performance is evaluated on actual dark matter models.
- Chapter 5 brings the results of this work. First, the motivation behind the choice of Segue 1 as the optimal dark matter candidate for searches with MAGIC is presented. This is followed by the details on the observations and the full likelihood analysis of the gathered data. Lastly, this chapter ends with interpretation of the achieved results in the light of different models of dark matter annihilation and decay.
- Chapter 6 gives a brief insight into the future of the Cherenkov astronomy and prospects for dark matter searches with the Cherenkov Telescope Array.

A short summary of the most relevant points of this thesis is presented in conclusions.

As an additional note, it should be mentioned that, at the time of publication of this book and two years after the thesis was written in 2013, work presented here still provides the most constraining limits to properties of dark matter particles, from observations of dwarf spheroidal galaxies with Cherenkov Telescopes. As for the full likelihood method, it is now becoming a standard in the field. In fact, it has already been used to calculate the prospects for dark matter searches in satellite galaxies and galaxy clusters with CTA (CTA Consortium, in preparation). Furthermore, instigated by this thesis, there is an ongoing effort to combine the results from MAGIC observations of Segue 1 with the results from 6 years of Fermi-LAT data of 15 dwarf spheroidals (published earlier in 2015). The combined analysis will set global, optimized, and strongest bounds on dark matter properties currently obtainable from gamma-ray astronomy (J. Aleksić et al., in preparation).

Barcelona
May 2015

Dr. Jelena Aleksić

Acknowledgments

I begin these acknowledgments by expressing my deepest gratitude to my supervisors, Manel Martinez and Javier Rico. Over the years I could always turn to you for advice, ideas, discussions, and support. I really appreciate all the efforts you have invested in my scientific formation. This thesis is the reflection of your work as it is of mine.

I am also grateful to Abelardo Moralejo, for all the help and advices. Also, to Juan Cortina, Oscar Blanch, Daniel Mazin, and Victor Stamatescu. A very special thanks goes to Stefan Klepser, for showing me many tips and tricks of the analysis. To Roger Firpo, for all the effort in meeting my computing requests. And in particular to Julian Sitarek, for his endless patience with my endless questions. I really appreciate it, guys.

I want to thank Alejandro Ibarra, for great collaboration and his whole TUM group, for the warm welcome and many cakes during my stay with them.

I have to acknowledge those MAGIC Collaboration colleagues I worked with and learned from. Particularly, I would like to thank my dark matter comrades, Michele Doro, Saverio Lombardi, Daniel Nieto, and Fabio Zandanel.

And I cannot forget all the IFAE students with whom I have shared many happy… They'll understand. Diego, Roberta, Ignasi, Gianluca, Manel E, Alicia, Adiv, and Ruben—it's been fun.

Veliko hvala mojoj porodici i mojim prijateljima na podršci. A todos mis amigos; especially during these past few months.

Aquest treball s'ha fet amb el suport del Departament d'Universitats, Recerca i Societat de la Informaci de la Generalitat de Catalunya. This work has been partially funded through the MICINN's MultiDark project CSD2009-00064.

Contents

Abbreviations

a.s.l.	above sea level
ACM	Active Mirror Control
ADC	Analog to Digital Converter
AMS	Alpha Magnetic Spectrometer
ATLAS	A Toroidal LHC Apparatus
BAO	Baryon Acoustic Oscillations
BBN	Big Bang Nucleosynthesis
BM	BenchMark
c.l.	confidence level
C.U.	Crab Nebula Unit
callisto	CALibrate Light Signals and Time Offsets
CC	Central Control
CCNN	Close Compact Next Neighbor
CDM	Cold Dark Matter
CMB	Cosmic Microwave Background
CMS	Compact Muon Solenoid
cMSSM	constrained Minimal Supersymmetric Standard Model
CoG	Center of Gravity
CTA	Cherenkov Telescope Array
DAQ	Data Acquisition System
DEIMOS	DEep Imaging Multi-Object Spectrograph
DRS	Domino Ring Sampler
dSph	dwarf Spheroidal (galaxy)
DT	Discriminator Threshold
EAS	Extensive Air Shower
EGRET	Energetic Gamma-Ray Experiment Telescope
EM	ElectroMagnetic
FITS	Flexible Image Transport System
FoV	Field of View
FSR	Final State Radiation
GRB	Gamma-Ray Burst

GST	(Fermi) Gamma-Ray Space Telescope
H.E.S.S.	High Energy Stereoscopic System
HDU	Header and Data Unit
HE	High Energy
IACT	Imaging Air Cherenkov Telescope
IVOA	International Virtual Observatory Alliance
KK	Kaluza–Klein
LAT	Large Area Telescope
LHC	Large Hadron Collider
LKP	Lightest Kaluza–Klein Particle
LSP	Lightest SUSY Particle
LUT	Look-Up Table
MACHO	MAssive Compact Halo Object
MAGIC	Major Atmospheric Gamma-ray Imaging Cherenkov (telescopes)
MARS	MAGIC Analysis and Reconstruction Software
MC	Monte Carlo
melibea	MErge and Link Image parameter Before Energy Analysis
merpp	MERging and Preprocessing Program
MOND	MOdified Newtonian Dynamics
MONSTER	Magic Optical NanoSecond Trigger and Event Receiver
MSSM	Minimal Supersymmetric Standard Model
mSUGRA	minimal SUper GRAvity
MUX	MUltipleX readout system
NFW	Navarro, Frank, and White
NSB	Night Sky Background
PAMELA	Payload for Antimatter Matter Exploration and Light-nuclei Astrophysics
PDF	Probability Density Function
ph.e.	photoelectron
PHDU	Primary Header and Data Unit
PMT	Photo-Multiplier Tubes
PSF	Point Spread Function
QE	Quantum Efficiency
RF	Random Forest
ROI	Region of Interest
SDSS	Sloan Digital Sky Survey
SED	Spectral Energy Distribution
SEGUE	Sloan Extension for Galaxy Understanding and Exploration
SM	Standard Model
sorcerer	Simple, Outright Raw Calibration; Easy, Reliable Extraction Routines
star	STandard Analysis and Reconstruction
SUSY	SUperSYmmetry
TeVeS	TEnsor–VEctor–Scalar gravity
UED	Universal Extra Dimensions

UFO	Unassociated Fermi Objects
VCSELs	Vertical Cavity SurfacE Laser diodes
VERITAS	Very Energetic Radiation Imaging Telescope Array System
VHE	Very High Energy
VIB	Virtual Internal Bremsstrahlung
VO	Virtual Observatory
WIMP	Weakly Interacting Massive Particle

List of Figures

List of Tables

Chapter 1
Dark Matter Paradigm

The notion of dark matter has been present for almost a century, but the question about its nature is still unanswered. Observational evidence and cosmological predictions assure that dark matter represents almost 85 % of the matter content of our Universe, and more than 25 % of its total energy budget. Discovering its essence is one of the most important and exciting tasks of modern science.

This chapter is devoted to a brief introduction of the dark matter concept as well as the experimental results and theoretical predictions that support this paradigm. Additionally, some of the most widely considered candidates for the dark matter particle are presented.

1.1 Observational Evidence

Although works of Öpik [1] and Oort [2] were implying that the luminous components of galaxies were not sufficiently massive to explain their dynamics, it was Zwicky that first adopted the concept of dark matter in 1933 [3]. Through studies of individual galaxies in the Coma cluster, he concluded that their radial velocities were too high for the system to be gravitationally stable and not disperse—unless there was a dominant, invisible, *missing mass* keeping it together.

In the decades that followed, more observational evidence supporting the dark matter paradigm was gathered on all scales. This section briefly describes some of the most relevant results.

1.1.1 Dynamics of Galaxies and Galaxy Clusters

Among the most compelling and direct evidence for the existence of dark matter are the rotation curves, i.e. circular velocities of visible stars and gas given as a function of their radial distances from the galaxy center.

J. Aleksić, *Optimized Dark Matter Searches in Deep Observations of Segue 1 with MAGIC*, Springer Theses, DOI 10.1007/978-3-319-23123-5_1

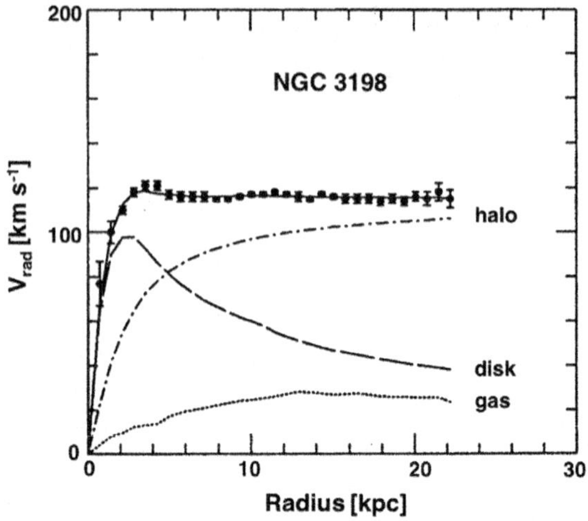

Fig. 1.1 Rotation curve of galaxy NGC 6503. Also shown are the velocity contributions from the visible baryonic components, gas (*dotted line*) and galaxy disk (*dashed line*), as well as the dark matter contribution (*dot-dashed line*) required to match the observations. Adapted from [5]

In a stably bound system with spherical distribution of mass objects (stars, galaxies, etc.), the virial theorem relates the total time-averaged kinetic and potential energies, so that velocity distribution of an element of the system is derived as $v(r) \propto (M(r)/r)^{1/2}$, where r is the distance from the center and $M(r)$ is the mass of the system within radius r. Following this reasoning, in the innermost regions of the elliptical galaxies and the bulges of the spheroidal ones, where mass distribution can be considered spherical and of constant density, $v(r) \propto r$ applies. As one moves towards the galaxy outskirts, the stars become scarce and very spread apart, $M(r)$ is approximately constant, and $v(r)$ is expected to decline as $r^{-1/2}$. The observations, however, show that circular velocities do not decrease with the increasing distance from the galactic center: instead, rotation curves show an unexpectedly flat behavior that extends far beyond the edges of the luminous components (Fig. 1.1, [4, 5]). This implies that mass distributions of galaxies cannot be as concentrated as their light distributions, i.e. *mass does not trace light.*

This departure from predictions of Newtonian gravity is apparent on greater scales as well: the mass of a galaxy cluster, estimated from the velocities of its constituting galaxies, is much larger than the mass seen directly as galaxies and hot gas in the cluster.

The discrepancy is solved if a dominant dark matter halo embedding the galaxies and galaxy clusters is added to the whole system. A common way to express the estimated amount of dark matter needed for the dynamical equilibrium is through the mass-to-light ratio (M/L), commonly given in solar units, $M_\odot/L_\odot$. The mass is estimated from the dynamics of the system and gravitational lensing (Sect. 1.1.2), while the light is inferred from the photometric observations. For spiral galaxies, for instance, M/L is of order of 10. On larger scales, for galaxy clusters, this ratio is of order of 100. The most dark matter dominated systems known so far are the dwarf spheroidal (dSph) galaxies, the low surface brightness objects that show unusual velocity dispersions and indicate a complete domination of dark over luminous matter at any radii (Sect. 5.1.1), with M/L reaching values of order of 1000.

1.1.2 Evidence from Gravitational Lensing

According to the predictions of General Relativity, gravitational fields deflect the light path of photons and alter the apparent flux and shape of astronomical sources. Mass concentrations modify the space-time metric and act as lenses, causing background objects to seem distorted to the observer. This effect can be used to ascertain the presence of mass even when it emits no light and, furthermore, even probe its nature to the certain extent [6]. Gravitational lensing is primarily sensitive to the amount of matter of the lensing body and, in accordance to its effects, distinction is made between the strong, weak and microlensing.

The weak lensing regime has proven to be particularly useful in probing the structure and the dynamical properties of dark matter halos. One of the most spectacular results from lensing effects, that not only favors the dark matter paradigm but sheds some light on its nature as well, is the case of the Bullet cluster of galaxies [7, 8]. Its baryonic distribution, observed in X-rays by Chandra, and gravitational mass, mapped from weak lensing, indicate that the Bullet cluster formed in a violent collision of two galaxy clusters. Two massive substructures, that are offset with respect to the baryon components, imply that the dark matter halos of the merging clusters passed through each other, succumbing only to weak and gravitational interactions, while the baryonic contents were slowed down by the electromagnetic force. As a result, the baryonic and dark matter portions of each of the clusters were decoupled (Fig. 1.2). These findings strongly favor the dark matter existence, seen how otherwise the cluster galaxy population would not be following, but coinciding with the mass distribution. Furthermore, this case also suggests the collisionless nature of dark matter particles.

Effects from other lensing regimes also contribute to our understanding of dark matter. For instance, measurements from strong lensing imply that galaxies and galaxy clusters are dark matter dominated, with M/L increasing with the mass as well as the radius [9]; however, these values are often overestimated and biased towards the uppermost M/L limit. As for the microlensing, its effects can play a role in constraining the contribution of the dark baryonic bodies (like dark stars and Massive Compact Halo Objects (MACHOs)) to the total dark matter content of the Universe. The current constraints from microlensing surveys, however, exclude MACHOs with masses in the $(6 \times 10^{-8} - 15)\ M_{\odot}$ range, setting the possible halo mass fraction of MACHOs to less then 8 % [10].

Fig. 1.2 Hubble Space Telescope and Chandra-XRO composite image of 1E0657-56 (the Bullet cluster), formed in collision of two galaxy clusters. The X-ray emitting gas, accounting for most of the baryonic matter of the system, is colored *red*, while the dark matter distribution, reconstructed from the gravitational lensing, is shown in *blue color*. *Credit* X-ray: NASA/CXC/CfA/ M. Markevitch et al.; Lensing map: NASA/STScI; ESO WFI; Magellan/U. Arizona/ D. Clowe et al.; Optical: NASA/STScI; Magellan/U. Arizona/D. Clowe et al.

1.2 Λ Cold Dark Matter Model

If the observational findings presented in Sect. 1.1 are justified by the existence of dark matter, a cosmology is needed that can successfully incorporate this dark matter in the evolution and current image of our Universe. The most valid and elegant explanation so far is offered by the Λ Cold Dark Matter (ΛCDM) model, a cosmology based on both sound theoretical foundations and good agreement with the observational results (for a review, see, e.g. [11] and references within).

The Big Bang Theory is one of the pillars supporting the ΛCDM: by extrapolating backwards in time the measurements of the energy content of the Universe today, at some 10^{10} years ago a singularity is reached—the so-called *Big Bang*. To be more precise, any trustworthy extrapolation breaks down before this singularity, since there are no known physics laws that can deal with such extreme conditions; thus, the Big Bang is defined as the very first stage of an extraordinarily hot and dense Universe.

The very beginning of the evolution of the Universe is still subject to the speculative ideas; the history of the Universe is established with more certainty, based on well understood and experimentally tested physics laws, from $t \simeq 10^{-10}$ s onwards. Some of the most relevant phases of the Universe evolution are:

- *Planck epoch*: the first 10^{-43} s, when the Universe was $\sim 10^{-60}$ of its current size, with temperature of $T \approx 10^{32}$ K. The current theories are still inadequate to explain what exactly was going on during this period.
- *Grand Unification epoch*: by its end ($t \simeq 10^{-36}$ s), previously unified forces separated: first the gravity from the gauge forces, and then the strong from the electroweak ones.
- *Inflationary epoch*: a period of accelerated expansion during which the size of the Universe increased by a factor of $\sim 10^{26}$. The growth was driven by the inflation field which, at the end of the epoch ($t \simeq 10^{-32}$ s), decayed into the Standard Model (SM) particles. Inflationary period explains why the Universe is so big and homogeneous.
- *Electroweak phase transition*: while expanding, the Universe was cooling down, and when the temperature dropped below 10^{15} K, the electromagnetic and weak forces separated, causing the previously created particles to acquire mass.
- *Quark-hadron transition*: at $t \simeq 10^{-5}$ s and $T \sim 10^{12}$ K, quarks bounded with gluons to form hadrons. The dark matter particles could have been created in this epoch.
- *Leptons era*: first the neutrinos froze out ($t \simeq 1$ s, $T \simeq 10^{10}$ K), followed by the efficient electron–positron annihilation ($t \simeq 5$ s, $T \simeq 5 \times 10^{9}$ K). This left the Universe radiation-dominated.
- *Big Bang nucleosynthesis* (BBN): around $T \simeq 10^{9}$ K ($t \sim 200$ s) the strong interactions became important, causing protons and neutrons to combine into light elements (D, He, Li). This went on until the temperature dropped to $T \sim 10^{8}$ K ($t \sim 10^{3}$ s), which was insufficient to sustain further reactions, freezing the abundances of the light atomic nuclei (until the star formation period). The successful prediction of relic abundances of these elements is one of the most striking evidence confirming the Big Bang theory.
- *Matter-radiation equality*: at $t \sim 10^{11}$ s and $T \sim 10^{4}$ K, the charged matter particles and photons were strongly coupled in the plasma; the density fluctuations, left by the inflationary epoch, began to grow in amplitude. According to ΛCDM, dark matter dominated at this stage.
- *Recombination*: some 4×10^{5} years after the Big Bang ($T \sim 10^{3}$ K), electrons combined with nuclei into electrically neutral atoms. The photons decoupled, forming the free-streaming cosmic microwave background (CMB); the Universe became transparent to light.
- *Dark ages*: from $t \sim (10^{5}$–$10^{8})$ years; no objects were formed yet.
- *Reionization*: at $t \approx 10^{8}$ years after the Big Bang, the small ripples in the matter density gradually assembled into stars and galaxies; their high energy photons began to ionize the surrounding H. Meanwhile, the most massive stars ran out of nuclear fuel and exploded as supernovae, with heavy elements (C, O,...) created as by-products.
- *Dark energy*: at $t \sim 10^{9}$ years, a negative pressure dark energy came to dominate the Universe, whose expansion started to accelerate.
- *Today*: $t = 13.7 \times 10^{9}$ years, $T = 2.7$ K.

ΛCDM is also founded on General Relativity and the following principles: Earth does not occupy a preferential place in the Universe (*Copernican principle*) and on the sufficiently large scales the Universe is homogeneous and isotropic (*Cosmological principle*). The most general space-time metric satisfying these conditions is the Friedmann–Robertson–Walker metric:

$$ds^2 = -c^2dt^2 + a(t)^2\left(\frac{dr^2}{1-kr^2} + r^2\left(d\theta^2 + \sin^2\theta d\phi^2\right)\right), \tag{1.1}$$

where (r, θ, ϕ) are the spherical coordinates; $a(t)$ is the cosmic scale factor and k is the spatial curvature constant, that can take values of 0, -1 and $+1$ for the flat, negatively curved and positively curved Universe geometry, respectively.

The dynamical evolution of the $a(t)$ for the Universe containing different components with total energy density ρ and relativistic pressure p, is determined by the General Relativity equations reduced to two independent Friedman equations:

$$H^2 \equiv \left(\frac{\dot{a}}{a}\right)^2 = \frac{8\pi G\rho}{3} - \frac{kc^2}{a^2} \tag{1.2}$$

and

$$\left(\frac{\dot{a}}{a}\right) = -\frac{4\pi G}{3c^2}(\rho + 3p), \tag{1.3}$$

where H is the Hubble parameter, defined as the expansion velocity of the Universe. In ΛCDM, the total ρ is divided among three main components: matter (baryonic and dark, $\rho_{\mathrm{m}} = \rho_{\mathrm{b}} + \rho_{\mathrm{dm}}$), radiation ($\rho_{\mathrm{r}}$) and dark energy ($\rho_{\Lambda}$):

$$\rho = \rho_{\mathrm{m}} + \rho_{\mathrm{r}} + \rho_{\Lambda}. \tag{1.4}$$

For a Universe with a spatially flat metric ($k = 0$), the total energy density from Eq. (1.2) becomes the *critical density*:

$$\rho_{\mathrm{c}} \equiv \frac{3}{8\pi G}H_0^2, \tag{1.5}$$

where H_0 is the Hubble constant, i.e. the value of the Hubble parameter today ($H_0 = 67.3 \pm 1.2\,\mathrm{km\,s^{-1}\,Mpc^{-1}}$ [12]). Normalized to the critical density, Eq. (1.4) becomes the *density parameter* Ω:

$$\Omega \equiv \frac{\rho}{\rho_{\mathrm{c}}} = \Omega_{\mathrm{m}} + \Omega_{\mathrm{r}} + \Omega_{\Lambda} + \Omega_k = 1, \tag{1.6}$$

where $\Omega_k \equiv -kc^2/(aH_0)^2$ is the curvature density contribution.

In general, the equation of state of each of the density constituents is defined by a proportionality constant w, such that $p = w\rho$. Specifically, at the present epoch

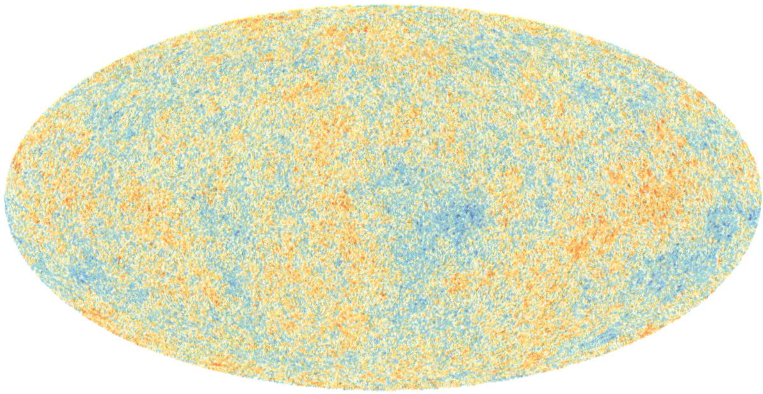

Fig. 1.3 The anisotropy map from the 15.5 months of measurements with the Planck space telescope, showing the temperature fluctuations in the early Universe. The covered temperature range is of $\pm 500\,\mu\text{K}$. Taken from [13]

($a = 1$), components contributing to the total energy density of the Universe are the non-relativistic matter ($w = 0$), radiation ($w = 1/3$) and dark energy ($w = -1$).

1.2.1 Cosmic Microwave Background

Probably the most convincing confirmation of ΛCDM comes from its success in predicting the initial candidates for structure formation, which culminated in the discovery of the temperature fluctuations in the CMB.

CMB represents a relic from the earliest stages of the evolution of the Universe. It is an almost black body (T = 2.725 K), isotropic radiation, composed of photons frozen-out at the mass-radiation decoupling era and cooled down by the expansion of the Universe. The experimental confirmation of CMB in 1965 [14] served as evidence for the validity of the Big Bang theory. Most recently, the predicted temperature anisotropies (at the scale of 10^{-5} K) have been measured by the Planck space telescope [15] with unprecedented sensitivity (Fig. 1.3, [13]). However, CMB is also expected to provide information on the Universe during the epoch of recombination: ΛCDM predicts that the acoustic waves—forming in the photon-baryon fluid as a result of conflict between the photon pressure and baryonic gravitational potential—froze when the photons decoupled. At the present epoch, those fluctuations should be seen as a series of peaks and troughs in the observed angular power spectrum, whose positions and amplitudes can be used to constrain a variety of cosmological parameters (Fig. 1.4). In particular, the location of the first peak probes the overall energy content of the Universe, while the relative heights of the peaks constrain the baryonic density Ω_b.

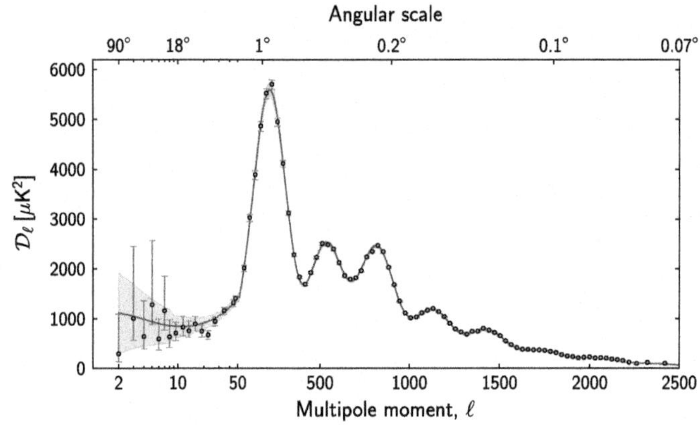

Fig. 1.4 The temperature angular power spectrum of the primary CMB from the Planck space telescope, showing the precise measurement of seven acoustic peaks, that are well fitted by the ΛCDM theoretical model (*full line*). Taken from [13]

These estimates of cosmological parameters are complemented by the values obtained from other techniques. By measuring the luminosity peaks of type Ia supernovae, together with their redshift z, constraints can be put on the time-evolution of the H parameter, which in turn gives bounds on the matter and radiation content of the Universe [16]. Additionally, precisely determined values of Ω_b are obtained from the BBN, as the relic abundances of the elements formed during this epoch (D, He, Li) are sensitive to the baryon density [17]. Baryon Acoustic Oscillations (BAO), marking the imprint of sound waves in the pre-recombination plasma, leave, at a very large scale ($\sim$100 Mpc), a subtle characteristic bump in the clustering of galaxies, whose properties are related to the evolution of the Universe [18].

The most precise determination of the different cosmological parameters is obtained by global analysis that beside the CMB anisotropies also includes the measurements from other techniques (Fig. 1.5). Currently the best-fit values of the density parameters are [12]:

$$\begin{aligned}\Omega_b &= 0.0486 \pm 0.002,\\ \Omega_{dm} &= 0.2647 \pm 0.015,\\ \Omega_\Lambda &= 0.6850^{+0.017}_{-0.016}.\end{aligned}$$

These values show that the Universe today is dominated by the dark energy, that drives its accelerated expansion and contributes to the ρ with $\sim$68.5 %. The remaining energy content is composed of matter, out of which almost 85 % is non-baryonic. Contribution of the relativistic particles is negligible to the total energy budget of the Universe today: photons and neutrinos only contribute with $\Omega_r < 0.1\,\%$.

For a more complete review of all cosmological parameters and their implications on our understanding of the Universe, refer to [12, 13].

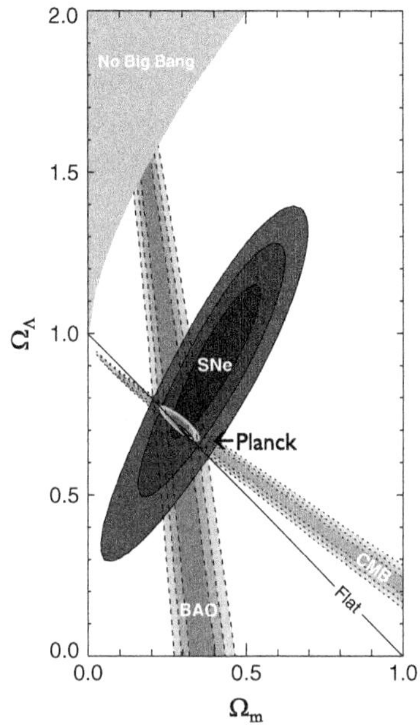

Fig. 1.5 Compilation of different bounds on the preferred region in the Ω_m–Ω_Λ plane [19]. The superimposed black contours are from the Planck measurements. Adapted from [12]

1.2.2 Large Structure Formation

Another relevant confirmation of ΛCDM comes from the remarkable agreement between the predictions of numerical simulations, based on ΛCDM, and the actual measurements, by redshift surveys, of the large scale structures of the Universe.

In the ΛCDM context, the structure formation initiated from the primordial perturbations in the gravitational potential, presumed to be seeded by quantum fluctuations in the inflation era (Sect. 1.2). As the Universe evolved from this initial inflationary state, regions with density higher than the average background density (expressed in terms of density contrast, $\delta(x) = (\rho(x) - \langle\rho\rangle)/\langle\rho\rangle)$ were able to grow due to the gravitational collapse. During the matter-dominated epoch, these tiny density fluctuations grew linearly, as $(1+z)^{-1}$, until $\delta(x) \sim 1$. The fluctuations then entered the non-linear regime, collapsed and the formation of the gravitationally bound structures began [20].

From the CMB temperature fluctuations (Sect. 1.2.1), it is known that the largest δ in baryonic matter at the time of recombination had an amplitude of $\sim 10^{-5}$; in a matter-dominated Universe, δ could not have grown more than a factor 10^3 since; thus, if the Universe was baryonic, the amplitude of fluctuations today would be $\delta < 0.1$. This is in sharp contrast with $\delta \gg 1$, needed to form the present-day structures. The existence of dark matter provides a simple solution to this problem: dark matter decouples while the bosons and fermions are still in the thermal equilibrium and strongly interacting. That way, the density fluctuations in dark matter begin growing earlier than in the ordinary matter. After the recombination, baryons collapse into the already existing gravitational wells formed by the dark matter overdensities, thus following the dark matter distribution. And given the overwhelming fraction of dark matter in the total mass budget of the Universe, the dark matter component stands out as the driving factor behind the structure formation.

And what can structure formation say about the nature of dark matter particles? For particles to form a structure, it is necessary for their free-streaming paths to be smaller than the fluctuation scale—otherwise, particles do not feel the gravitational pull from the fluctuations and can freely scatter, diluting the density and preventing any structure formation at the given scale. Based on the effect they have on the structure formation, three different types of particle dark matter have been postulated: hot, cold and warm dark matter.

- *Hot dark matter* particles are expected to have very large free-streaming lengths, due to their relativistic nature at the freeze-out. This implies that density fluctuations below the Mpc scales would be erased. The supercluster-scale structures would have to form first, and then fragment into smaller objects. This, so-called *top-down* scenario, is strongly disfavored by the observations of high-redshift galaxies. Therefore, hot dark matter can not be the dominant dark matter type [21].
- *Cold dark matter* particles would be non-relativistic with small free-streaming lengths, thus allowing formation of Earth-like masses. Such particles support the hierarchical structure formation—from smaller to larger sizes (the *bottom–up* scenario). Cold dark matter hypothesis is favored by the excellent agreement between the observational data and simulations [22].
- *Warm dark matter* particles are the intermediate solution between the hot and cold dark matter. Their free-streaming lengths are of galaxy sizes [23], suggesting bottom–up formation at larger scales, and top–down scenario at smaller ones.

As already mentioned, the validity of the cosmological models can be tested through the N-body simulations. These numerical studies attempt to reproduce the current image of the Universe at large scales, as well as the formation of galactic halos for chosen cosmological circumstances. So far, the best agreement between the simulations and measurements is achieved for the ΛCDM model, with cold and weakly interacting dark matter (Fig. 1.6). N-body simulations are also used to generate mock galaxy catalogues and maps that yield the observed correlations and clustering of galaxies, and precise values of the cosmological parameters, in combination with the CMB maps and other cosmological probes.

1.2.3 Challenges to the ΛCDM

Despite the exceptional agreement between ΛCDM and majority of the cosmological data, there are still some observational results that can not be accurately justified by the model (see, e.g. [28] and references within). Details on few of the most striking issues challenging the ΛCDM concept are presented below.

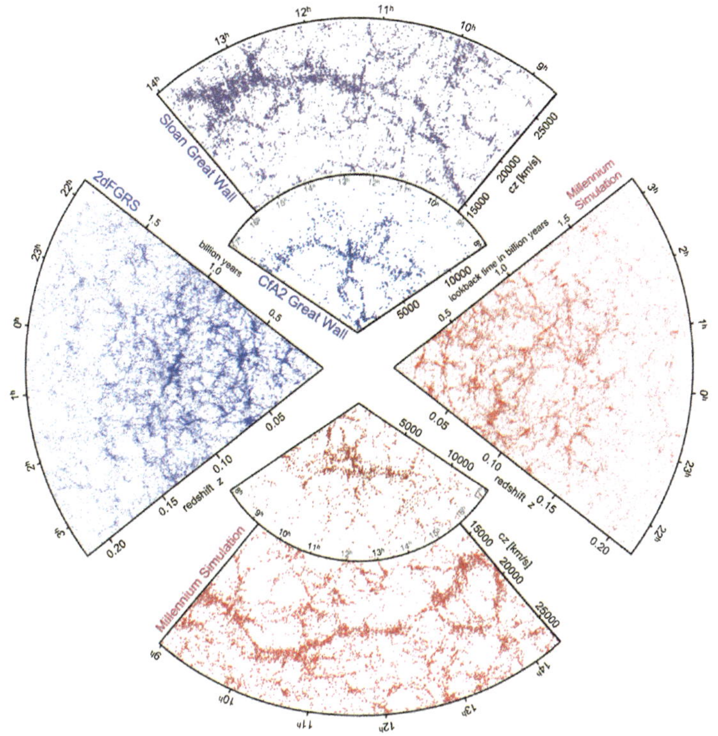

Fig. 1.6 The large scale structure of the Universe, mapped by the 2dF Galaxy Redshift Survey [24], the Sloan Digital Sky Survey [25] and the CfA Redshift Survey [26]. Each point represents a galaxy as a function of right ascension and redshift. Also shown are the corresponding N-body Millennium simulations [27] of the ΛCDM structure formation. *Credit* VIRGO/Millennium Simulation Project

Dark Matter Halos Dark matter-dominated halos of galaxies are considered to be another generic success of ΛCDM, as mapped out by rotation curves (Sect. 1.1.1). However, the detailed predicted properties of halos do not seem to be well matched to observations in the inner regions of many disk galaxies [29]. The expected dark matter cusps (density $\rho \propto r^{-\alpha}$, with $1 < \alpha < 1.5$) are not found in most low surface brightness dwarfs observed at high resolution (Sect. 2.3.2.1, [30]). Whether this issue can be understood by more detailed, better resolution simulations that include the full interactions between the baryons and dark matter, remains to be seen. At the moment, theory fails to provide a compelling explanation.

The Missing Satellite Problem N-body simulations suggest that, in the hierarchical structure formation, dark matter halos are created clumpy, with substructure persisting through the successive mergers. One consequence of the substructure is a large population of satellite galaxies, approximately increasing in abundance with decreasing mass. That, however, contradicts the observations, as the number of detected dwarf galaxies is within an order of magnitude lower than expected from simulations [31]. This problem has two possible solutions. One is that the smaller dark matter halos do exist, but are simply not massive enough to have attracted sufficient baryonic matter and are therefore almost optically invisible. Indeed, in the last decade a number of ultra-faint dwarfs has been discovered, with $M/L \sim 1000$ (Sect. 5.1.1, [32]). Other solution involves complex interactions between dwarfs and large galaxies, with dwarfs ending tidally stripped apart and extremely difficult to identify.

The Tully–Fisher Relation The baryonic Tully–Fisher relation is an observed dependency between the asymptotic circular velocity and baryonic mass in galaxies. Over the range of five decades in baryonic mass, this relation has remarkably little intrinsic scatter and is well described by a simple power law of slope $\alpha \approx 4$. This, however, cannot be matched by the galaxies assembled inside the ΛCDM halos, for which the simplest predictions give a slope of $\alpha \sim 3$ [33].

Among other challenging discrepancies are the large scale velocity flows, the low multipoles in the CMB, the quasars optical alignment and the imperfect fit to the rotation curves. However, whether the solution of these issues lies in the domain of Fundamental Gravity, Particle Physics or Astrophysics, these problems are not sufficient to abandon the ΛCDM concept. Rather, at this point it can be concluded that ΛCDM successfully matches the Universe on large scales, while on small scales it has (possibly) some difficulties in confronting the observations.

1.2.4 Alternative Cosmologies

The above mentioned shortcomings of ΛCDM have motivated the development of alternative cosmological explanations. Based on the assumptions that dark matter is not cold or weakly interacting, or even postulating that dark matter does not exist at all, the majority of these alternative cosmologies provide solutions to singular problems, while at the same time create new inconsistencies. Until date, neither of the proposed cosmologies offers an evolutionary image of the Universe as completely as ΛCDM does. However, for the sake of argument, the best-justified of these alternatives are briefly described.

Modified Newtonian Dynamics (MOND) claims that the law of gravity deviates from the Newtonian one, thus eliminating the need for existence of dark matter [34]. According to MOND, below a certain gravity scale the effects of the gravitational force are magnified. This would explain the observed flattening of the rotation curves, as well as the Tully–Fisher relation. However, MOND fails to explain the dynamics of large objects like galaxy clusters, as well as the gravitational lensing effects without adding an additional component of the matter. More over, MOND can not account for any relativistic phenomena, and overall, does not provide a satisfactory cosmology.

Tensor–Vector–Scalar (TeVeS) Gravity is developed as the relativistic generalization of MOND [35]. TeVeS works in the weak-field limit and possesses all good qualities of the MOND theory. In addition, TeVeS can explain gravitational lensing effects, although in a way non-consistent with the galaxy rotation curves. Major drawbacks of the TeVeS gravity model are the incompatibility with stellar evolution theory and the inability to explain the Bullet cluster phenomena (Sect. 1.1.2).

Other non-standard cosmological models include the Gödel Universe (homogeneous, rotating Universe in which closed time-like curves exist, [36]), the Quasi-Steady state cosmology (the expanding Universe with no beginning and no end, with new matter continuously created, [37]), the Einstein-de Sitter Universe (flat, matter dominated Universe of infinite total volume, [38]), etc.

1.3 Dark Matter Candidates

Despite the overwhelming observational evidence for the existence of dark matter, the nature of its constituent is still unknown. The wealth of recent astrophysical and cosmological data, however, imposes significant constraints on the dark matter properties. In context of ΛCDM, the dark matter particle should:

- be neutral—otherwise, it could couple with photons and therefore would not be dark;
- match the dark matter relic density, $\Omega_{\rm dm}h^2 = 0.1198 \pm 0.0026$[1];
- be stable on cosmological scales, so that it was present in the early Universe and is still around today;
- interact only weakly and gravitationally: the couplings with electromagnetic sector, as well as strong interactions are highly suppressed by the observations;
- play a leading role in the structure formation in the Universe, as the fluctuations in the dark matter density are dominating the evolution of the perturbations in the matter-dominated era;
- be consistent with the BBN and not contradict the observed abundances of light elements;
- not affect the stellar evolution;
- be experimentally verifiable and consistent with the constraints derived by different methods of dark matter searches (Chap. 2).

This section presents some of the theoretically best-motivated dark matter particle candidates.

1.3.1 Weakly Interacting Massive Particles

Weakly interacting massive particles (WIMPs) are the most studied dark matter candidates. WIMPs are postulated as non-baryonic, stable and weakly interacting,

[1]In the multicomponent dark matter scenarios, the relic density for a given particle can be lower than the total $\Omega_{\rm dm}h^2$.

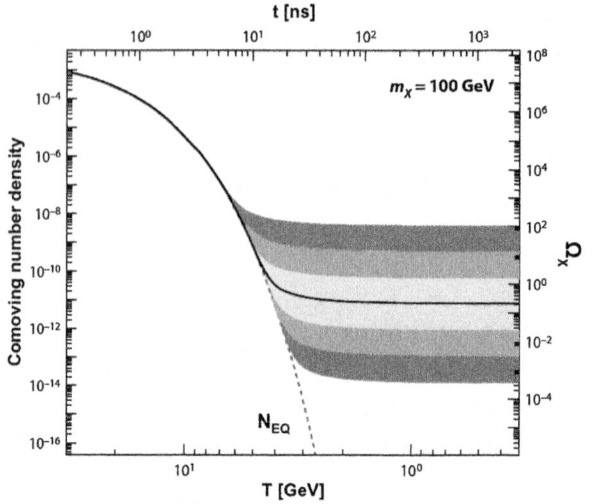

Fig. 1.7 The comoving number density (*left*) and resulting thermal relic density (*right*) of a 100 GeV annihilating dark matter particle as a function of temperature (*bottom*) and time (*top*). The *solid black line* corresponds to annihilation cross section that yields the correct relic density, and the *yellow*, *orange* and *violet regions* are for cross sections that differ by a factor 10, 100 and 1000 (respectively) from this value. The *dashed gray line* is the number density of a particle that remains in thermal equilibrium. Adapted from [39]. (Color in online)

but their greatest appeal is that they are found in many particle physic theories, that they naturally produce the correct relic density, and that they can be detected in many ways [40, 41].

WIMPs are presumed to be the relics of the Big Bang (Sect. 1.2). In the early Universe, WIMPs were in the thermal equilibrium with the primordial plasma; when temperature dropped bellow the WIMP mass m_χ, these particles decoupled, their production ceased and their number density began to drop exponentially, as $e^{-m_\chi/T}$ (Fig. 1.7). However, as the Universe kept expanding, the dark matter gas became so diluted, that WIMPs could no longer find each other to annihilate. At that point, WIMPs density froze-out, and their number asymptotically approached a constant—their thermal relic density.

The fact that this relic abundance is compatible with the estimate of the present dark matter density is known as the *WIMP miracle*. That is, the relic density today, given as

$$\Omega_\chi h^2 \approx \frac{10^{-27}\ \mathrm{cm}^3\ \mathrm{s}^{-1}}{\langle\sigma_{\mathrm{ann}} v\rangle}, \tag{1.7}$$

where Ω_χ is the WIMP density parameter, h is the scaled Hubble constant and $\langle\sigma_{\mathrm{ann}} v\rangle$ refers to the thermally averaged product of the annihilation cross section and velocity, is naturally produced by a thermal relic with a mass and interaction cross section on the weak scale.

For WIMPs the freeze-out occurred at $T \simeq m_\chi/20$, which sets the m_χ value in a few GeV—few TeV range. Therefore, at the time of decoupling, these particles were non-relativistic, and as such are suitable candidates for the role of cold dark matter. Additionally, as the WIMP number density froze before the BBN epoch, WIMPs are the oldest remnants: if discovered, they would give, for the first time, information on the very earliest stages of the Universe.

However, such particle—a WIMP—does not exist within the SM framework. Hence, one must go beyond the borders of the SM and into the realm of some new physics to try to find a WIMP. Some of these new theories and their WIMP candidates are described below. Accent is set on those particles that are of relevance for this work; for the others, references for further reading are provided.

1.3.1.1 Supersymmetric Dark Matter

Supersymmetry (SUSY) postulates that, for every SM particle there is a new, as-yet-undiscovered partner particle, with the same set of quantum numbers and gauge interactions but the spin, which is increased by 1/2. SUSY represents an elegant, theoretically sound scenario that relates fermions and bosons, thus creating a more unified picture of the particle world (see, e.g. [42, 43]). More importantly, SUSY provides possible solutions to some of the burning problems of the SM, like:

- *hierarchy problem*, linked to the vast discrepancy between the aspects of the weak nuclear force and gravity, is stabilized with SUSY through cancellation of divergences in the radiative corrections to the Higgs boson mass [44];
- *unification of the gauge couplings* of the SM is possible if the SUSY particles (sparticles) are included in the renormalization-group equations [45];
- *natural dark matter candidate* is provided by SUSY: the lightest SUSY particle (LSP) is expected to be heavy, neutral and stable, thus fitting perfectly in the ΛCDM paradigm as the dark matter constituent [46].

SUSY Realisations A general SUSY extension contains many unknown parameters. To make progress, it is typical to consider specific models in which simplifying assumptions unify many parameters, and then study to what extent the conclusions may be generalized.

The *Minimal Supersymmetric Standard Model* (MSSM) [47] is minimal is the sense that it contains the smallest possible field contents necessary to give rise to all the fields of the SM. The MSSM requires doubling of the SM degrees of freedom and two complex Higgs doublets, to account for the u and d quark masses, plus the complete set of SUSY partners (Table 1.1).

Since no sparticle with the same mass as its SM partner has been seen by the accelerator experiments (Sect. 2.1), SUSY has to be broken. Breaking of the symmetry, on the other hand, produces some critical effects on the proton lifetime, shortening it down to values lower than the age of the Universe, which contradicts the observations. To remedy the issue, a new discrete symmetry, called *R-parity*, is introduced. R-parity is defined as $R \equiv (-1)^{3B+2L+2s}$, where B, L and s stand for the baryon,

Table 1.1 Standard model particles and their superpartners in the MSSM model

SM particles/fields		SUSY partners			
		Interaction eigenstates		Mass eigenstates	
Quark	q	Squark	$\tilde{q}_L, \tilde{q}_R$	Squark	$\tilde{q}_1, \tilde{q}_2$
Lepton	l	Slepton	$\tilde{l}_L, \tilde{l}_R$	Slepton	$\tilde{l}_1, \tilde{l}_2$
Neutrino	ν	Sneutrino	$\tilde{\nu}$	Sneutrino	$\tilde{\nu}$
Gluon	g	Gluino	$\tilde{g}$	Gluino	$\tilde{g}$
W-boson	$W^{\pm}$	Wino	$\widetilde{W}^{\pm}$	Chargino	$\tilde{\chi}^{\pm}_{1,2}$
Higgs boson	$H^{\pm}$	Higgsino	$\widetilde{H}^{\pm}_{1,2}$		
B-field	B	Bino	$\widetilde{B}$	Neutralino	$\tilde{\chi}^0_{1,2,3,4}$
Higgs boson	$H^0_{1,2,3}$	Higgsino	$\widetilde{H}^0_{1,2}$		
W^3-field	W^3	Wino	$\widetilde{W}^3$		

lepton and the spin number, respectively. SM particles have R-Parity of +1, and their SUSY partners of -1. If R-parity is conserved, the sparticles can only be produced/annihilated in pairs, so that the LSP is stable and a viable dark matter candidate.

The main drawback of MSSM (in terms of practicality) is that has over 120 free parameters. The usual approach when studying SUSY effects is to assume a specific frameworks, where further well-motivated assumptions are introduced.

The *constrained Minimal Supersymmetric Standard Model* (cMSSM, [48]) is often regarded as the most simple and economical SUSY model. It reduces the number of needed parameters from 120 to only 5, by assuming certain universalities at the grand unification scale (where the SUSY breaking is expected to occur): all scalar particles have the same mass m_0, all the gauginos have the same mass $m_{1/2}$ and all the trilinear couplings A_0 are the same. The fourth continuous parameter is the vacuum expectation of the two Higgs fields, $\tan\beta$; the fifth parameter is the sign of the higgsino mass term μ, and it can be either positive or negative. The string of negative results in searches for cMSSM in collider experiments, however, puts this model on the verge of exclusion [49].

The *minimal Supergravity model* (mSUGRA, [50]) is a special case of the cMSSM: if the SUSY breaking is mediated by the gravitational effects, and the gravitino mass is fixed to the universal scalar mass ($m_{3/2} = m_0$), the cMSSM shrinks to the mSUGRA, with the same free parameters but in a reduced hyperspace where those parameters can move in (the μ parameter is no longer free, but fixed by demanding the radiative breaking of the electroweak symmetry).

Other MSSM realizations include the *non-universal Higgs masses model* (the SUSY breaking contributions to the Higgs masses do not have to be universal, [51]), *phenomenological MSSM* (based on phenomenology rather than on particular theoretical assumptions, with number of free parameters reduced to 19, [52]), etc.

SUSY Dark Matter Candidates Among the new particles that SUSY introduces, the electrically neutral ones with weak interactions are the natural dark matter candidates: the spin 3/2 fermion *gravitino* ($\widetilde{G}$), the spin 1/2 fermions called *neutralinos* ($\tilde{\chi}_1$, $\tilde{\chi}_2$, $\tilde{\chi}_3$, $\tilde{\chi}_4$), and the spin 0 scalars *sneutrinos* ($\widetilde{\nu}_e$, $\widetilde{\nu}_\mu$, $\widetilde{\nu}_\tau$).

The sneutrinos are not good dark matter candidates, as both their annihilation and scattering cross sections are large, so they are underabundant or excluded by null results from direct detection experiments, for all masses near m_{weak}. Gravitinos qualify as the dark matter particle in some particular scenarios, like the gauge mediated SUSY, where they are stable and lightest particles. Although theoretically well-motivated, $\widetilde{G}$ may be difficult to detect: as it interacts only gravitationally, the main source of gravitinos would be the decay of the next-to-lightest SUSY particles. Thus, the dark matter characterization in the gravitino scenario would require detection of the signatures of the progenitor particle.

Neutralinos, on the other hand, are favored as dark matter constituents. Neutralinos are mass eigenstates produced in mixing of the neutral, spin 1/2 fermions: bino ($\widetilde{B}$), wino ($\widetilde{W}_3$) and two higgsinos ($\widetilde{H}_1^0$ and $\widetilde{H}_2^0$) [46]. The lightest of the four resulting particles $\tilde{\chi}_1^0$ (henceforth just the neutralino χ), is expressed as

$$\chi \equiv \tilde{\chi}_1^0 = N_{11}\widetilde{B} + N_{12}\widetilde{W}^3 + N_{13}\widetilde{H}_1^0 + N_{14}\widetilde{H}_2^0, \tag{1.8}$$

where the coefficients N_{ij} are obtained by diagonalizing the neutralino mass matrix.

With R-parity conserved, the neutralino is stable LSP and thus an natural dark matter candidate, with relic density compatible with bounds from the Planck satellite (Sect. 1.2.1), a mass at the GeV–TeV scale, and a typical cross section of the order of weak interactions. Being a Majorana fermion, the neutralino can self-annihilate into (detectable) SM particles, such are:

- *Fermions:* the leading neutralino annihilation channels are into fermion pairs at tree-level, via s-channel through the exchange of Z or Higgs bosons, or via t-channel through sfermion exchange. The dominant final states are composed by heavy particles, like $\tau^+\tau^-$, $b\bar{b}$ and $t\bar{t}$ (for sufficiently high masses).
- *Photons:* direct annihilation into photons can occur at one loop level, as $\chi\chi \rightarrow \gamma X$, where $X = \gamma, Z$ or h. Such process is strongly suppressed, but not impossible; photons produced this way would be detected as sharp lines at energies $E_\gamma \approx m_\chi(1 - m_X^2/4m_\chi^2)$ (Sect. 2.3.2.3), representing an undoubtable evidence of dark matter detection. Photons can also be produced in the so-called internal bremsstrahlung [53]: if neutralinos annihilate into leptons, the annihilation exchange particle is a charged sparticle that can emit a photon. This photon restores the helicity in the annihilation processes of type $\chi\chi \rightarrow l^+l^-\gamma$, thus allowing for otherwise forbidden interactions. Photons produced this way are expected to carry a significant amount of energy ($E_\gamma > 0.6m_\chi$) and to produce a characteristic bump at the end of the differential photon energy spectrum (Sect. 2.3.2.3).
- *Gauge bosons:* in the low-velocity regime, pure gaugino-like neutralinos can annihilate into Z and $W^\pm$ bosons via t-channel, while pure higgsino-like and mixed neutralinos would produce these particles via s-channel.
- *Higgs boson:* neutralinos can annihilate into pair of Higgs bosons or a Higgs and a gauge bosons. The most favored channels are the annihilation into light

neutral Higgs and a Z boson ($\chi\chi \rightarrow h^0 Z$), into a heavy Higgs and a Z boson ($\chi\chi \rightarrow H^0 Z$), into a charged Higgs and a W boson ($\chi\chi \rightarrow H^{\pm}W^{\pm}$), and into a light Higgs and a pseudoscalar Higgs ($\chi\chi \rightarrow h^0 A_0$).

Details on methods and current progress of the searches for SUSY dark matter will be presented in Chap. 5. However, it should be mentioned that the latest results from Large Hadron Collider (LHC) [54] have struck serious blows to the SUSY credibility. No proof of new physics has been found so far. The newly discovered Higgs boson [55] behaves pretty much as the SM predicts, while none of the proposed SM extensions claims a completely "standard" Higgs. Furthermore, no hint of any anomalous behavior was detected in the extremely rare B_s meson decay [56]. Still, not all hope for the SUSY is lost: introduction of new parameters can adjust the model (to a certain extent) to the current experimental constraints. More conclusive results should be expected from 2015 on, when LHC begins to operate at the higher energy regime [57] (up to 14 TeV, compared to the current 8 TeV limit).

1.3.1.2 Universal Extra Dimensions

An alternative possibility for the new weak-scale physics are extra dimensions. The idea originated from work of Kaluza and Klein almost a century ago [58, 59], and since then it acquired many modern descendants, of which the theory of Universal Extra Dimensions (UED) has the strongest foundations [60].

In UED, all particles propagate in flat, compact extra dimensions of size 10^{-18} m or smaller. Every SM particle has an infinite number of partner particles of the same spin, with one at every Kaluza–Klein (KK) level n with mass $\sim nR^{-1}$, with R referring to the compactified radius of the extra dimension. Unlike SUSY, UED do not solve the gauge hierarchy problem; in fact, their couplings become large and nonperturbative at energies far below the Planck scale. However, UED are considered as plausible models under the assumption that they are a low-energy approximation to a more complete theory that resolves the hierarchy problem and is well-defined up to the Planck scale.

The simplest UED models preserve a discrete, KK parity, which implies that the lightest KK particle (LKP) is stable and a possible dark matter candidate. The LKP is typically a B^1, a level 1 partner of the hypercharge gauge boson. Investigations of the B^1 regions with the correct thermal relic density indicate that the required LKP mass is in the 600 GeV–1.4 TeV range, a slightly heavier than for the neutralino. Other possibilities for the LKP are KK neutron, KK Z and KK Higgs.

1.3.1.3 Other WIMP Candidates

Neutralino can be considered as a prototype of a WIMP, and KK dark matter an instructive alternative. There are many other examples, however. Some of the electroweak theories and their accompanying WIMP candidates include large extra

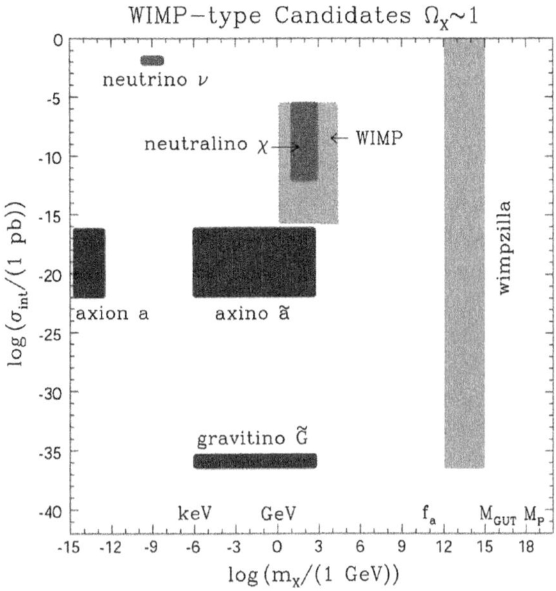

Fig. 1.8 Schematic representation of some well-motivated DM candidate particles. σ_{int} represents a typical order of magnitude of the interaction strength with the ordinary matter. The *box* marked as "WIMP" stands for several possible candidates. Taken from [68]

dimensions with branons as particles [61], little Higgs theories with T-odd particles [62] and warped extra dimensions with excited states [63]. As with all WIMPs, these dark matter candidates are produced through thermal freeze-out and are cold and collisionless, but their implications for detection may differ significantly.

1.3.2 Non-WIMP Dark Matter

As already mentioned, there is a vast collection of dark matter candidates (Fig. 1.8). As this work is focused on the search for WIMP particles, some of these other candidates are only briefly reviewed in this section.

1.3.2.1 Sterile Neutrinos

The fact that neutrinos have non-zero mass is a solid experimental evidence for new physics beyond the SM. For the neutrinos to get mass through the same mechanisms that generate masses for quarks and charged leptons, a new, right-handed neutrinos should be added. These, so-called *sterile neutrinos*, are weakly-interacting Majorana fermions. The lightest of sterile neutrinos, with mass predicted to be in the keV range, is compatible with warm dark matter (Sect. 1.2.2), but could also be considered as cold candidate, depending on the production mechanism. Additionally, sterile neutrinos may provide solution for the baryon asymmetry [64] and, in the warm dark matter scenario, the missing satellites problem [65].

As a consequence of the mixing between the sterile and active neutrinos (ν_s and ν, respectively), the former become unstable. The main decay channel for the lightest sterile neutrino is $\nu_s \to 3\nu$; from there, assuming the ν_s mass m_{ν_s} of order of 1 keV, the expected lifetime of the sterile neutrinos is estimated to 10^{17} years, meaning that these particles are cosmologically stable [66]. Through the described decay, the sterile neutrinos are extremely difficult to detect, given the low energy of the resulting active neutrinos. An alternative way of detection could be pursued for a subdominant decay channel that can provide a distinctive photon line, $\nu_s \to \gamma\nu$. Such line would contribute at energies of $E_\gamma = m_{\nu_s}/2$, and it would be broadened due to the velocity dispersion of sterile neutrino population. Therefore, compact regions with significant accumulations of sterile neutrinos could produce a detectable X-ray flux line in the 0.1–100 keV energy range.

1.3.2.2 SuperWIMPs

In the superWIMP framework for dark matter, WIMPs freeze-out in the early Universe (as described in Sect. 1.3.1), but later decay to superWIMPs, particles that form the dark matter that exists today [67]. Because superWIMPs are very weakly interacting, they have no impact on WIMP freeze-out, and the WIMPs decouple with a thermal relic density $\Omega_\chi \sim \Omega_{\rm dm}$. Assuming that each WIMP decay produces one superWIMP, the relic density of superWIMP is

$$\Omega_{s\chi} = \frac{m_{s\chi}}{m_\chi}\Omega_\chi. \tag{1.9}$$

SuperWIMPs therefore inherit their relic density from WIMPs, and for $m_{s\chi} \sim m_\chi$, the WIMP miracle also applies and the superWIMPs are produced in the desired amount to constitute much or all of dark matter.

Because the superWIMPs are very weakly interacting, they cannot be detected in conventional direct and indirect dark matter searches (Sects. 2.2 and 2.3). However, their extraordinarily weak couplings suggest that the decays of WIMPs to superWIMPs may have occurred very late, producing an observable impact on the BBN, the Planckian spectrum of the CMB, small-scale structure and the diffuse photon spectrum, thus possibly providing a way for superWIMP detection.

The superWIMP scenario is realized in many particle physics models. The prototypical example is gravitino, which exists in all SUSY theories. Other examples of superWIMP dark matter candidates include KK gravitinos in UED scenarios, axinos and quintessinos in SUSY theories, and many other.

1.3.2.3 Axions

The axion particle was proposed as a solution to the strong CP problem of the SM [69]. It is a light, neutral and weakly interacting, spin 0 boson that represents a natural dark matter candidate. For axions to live longer than the age of the Universe, their mass cannot exceed $m_a \leq 20$ eV. Axions can be produced thermally, like light gravitinos and sterile neutrinos, and in that scenario they would be the hot dark matter (Sect. 1.2.2). However, in order to achieve the correct relic density, the axion mass would have to be $m_a \sim 80$ eV; such particles can not be the bulk of the dark matter. The alternative is a non-thermal axion production, consequential to the spontaneous Peccei–Quinn (PQ) symmetry breaking. In that case, the axion can be a viable, cold dark matter particle, with mass in the 10^{-6}–10^{-2} eV range, and with properties dependent on whether the PQ phase transition occurred before or after the inflation epoch.

In the presence of electromagnetic fields, axions are predicted to couple with photons, leading to the so-called photon–axion oscillations. This effect may imprint a distinctive signature in the observed spectra of distant gamma-ray sources.

1.3.2.4 Hidden Dark Matter

Hidden dark matter is postulated as dark matter that has no SM gauge interactions [70]. Hidden sector differs from the visible one, as it can have its own matter content and gauge forces. Therefore, for the correct relic density to be achieved, for the hidden dark matter the WIMP miracle is generalized to the *WIMPless miracle*: dark matter naturally has the correct relic density, but does not necessarily have a weak-scale mass or weak interactions. Matter in hidden sector interacts only gravitationally. Still, in the scenarios that involve the existence of connector sectors, that mediate between the SM and the hidden sector, certain (detectable) astrophysical implications are possible.

References

1. E. Öpik, An estimate of the distance of the Andromeda nebula. Astrophys. J. **55**, 406–410 (1922)
2. J.H. Oort, The force exerted by the stellar system in the adrection perpendicular to the galactic plane and some related problems. Bull. Astron. Inst. Neth. **6**, 249–287 (1932)
3. F. Zwicky, Die Rotverschieb ung von extragalaktischen Nebeln. Helv. Phys. Acta **6**, 110–127 (1933)
4. V. Rubin, W.K. Ford Jr., Rotation of the Andromeda nebula from a spectroscopic survey of emission regions. Astrophys. J. **159**, 397–403 (1970)
5. K.G. Begeman, A.H. Broeils, R.H. Sanders, Extended rotation curves of spiral galaxies—dark haloes and modified dynamics. Mon. Not. R. Astron. Soc. **149**, 523–537 (1991)
6. P. Schneider, J. Ehlers, E.E. Falco, *Gravitational Lenses*, Study ed. (Springer, Berlin, 1992). ISBN: 978-3540665069

7. M. Bradač et al., Strong and weak lensing united III: measuring the mass distribution of the merging galaxy cluster 1E0657-56. Astrophys. J. **652**, 937–947 (2006), arXiv:astro-ph/0608408
8. D. Clowe, A direct empirical proof of the existence of dark matter. Astrophys. J. Lett. **648**, L109–L113 (2006), arXiv:astro-ph/0608407
9. R. Gavazzi et al., The Sloan Lens ACS survey. IV: the mass density profile of early-type galaxies out to 100 effective radii. Astrophys. J. **667**, 176–190 (2007), arXiv:astroph/0701589
10. P. Tisserand et al., Limits on the macho content of the galactic halo from the EROS-2 survey of the Magellanic Clouds. A&A **469**, 387–404 (2007), arXiv:astro-ph/0607207
11. S. Weinberg, *Cosmology*, 1st edn. (Oxford University Press Inc., New York, 2008). ISBN: 978-0198526827
12. P.A.R. Ade et al., Planck 2013 Results. XVI. Cosmological Parameters (2013), arXiv:1303.5076
13. P.A.R. Ade et al., Planck 2013 results. I. Overview of products and scientific results (2013), arXiv:1303.5062
14. A.A. Penzias, R.W. Wilson, A measurement of excess antenna temperature at 4080Mc/s. Astrophys. J. **142**, 419–421 (1965)
15. European Space Agency: Planck home page, http://sci.esa.int/science-e/www/area/index.cfm?fareaid=17 (2013)
16. M. Hicken et al., Improved dark energy constraints from ~100 new CfA supernova Type Ia light curves. Astrophys. J. **700**, 1097–1140 (2009), arXiv:0901.4804
17. G. Steigman, Primordial nucleosynthesis in the precision cosmology era. Annu. Rev. Nucl. Part. Sci. **57**, 463–491 (2007), arXiv:0712.1100
18. W.J. Percival et al., Baryon acoustic oscillations in the Sloan Digital Sky Survey Data Release 7 galaxy sample. Mon. Not. R. Astron. Soc. **401**, 2148–2168 (2010), arXiv:0907.1660
19. M. Kowalski et al., Improved cosmological constraints from new, old and combined supernova datasets. Astrophys. J. **686**, 749–778 (2008), arXiv:0804.4142
20. S. Dodelson, *Modern Cosmology*, 1st edn. (Academic Press, San Diego, 2003). ISBN: 978-0122191411
21. J.R. Primack, Whatever happened to hot dark matter? SLAC Beam Line **31**, 50–57 (2001), arXiv:astro-ph/0112336
22. V. Springel, C.S. Frenk, S.D.M. White, The large-scale structure of the Universe. Nature **440**, 1137–1144 (2006), arXiv:astro-ph/0604561
23. P. Colin, V. Avila-Reese, O. Valenzuela, Substructure and halo density profiles in a warm dark matter cosmology. Astrophys. J. **542**, 622–630 (2000), arXiv:astro-ph/0004115
24. The 2dF Galaxy Redshift Survey Collaboration: The 2dF Galaxy Redshift Survey home page, http://magnum.anu.edu.au/~TDFgg (2013)
25. The Sloan Digital Sky Survey Collaboration: Sloan Digital Sky Survey home page, http://www.sdss.org (2013)
26. J. Huchra et al., A survey of galaxy redshifts. IV—the data. Astrophys. J. Supp. **52**, 89–119 (1983)
27. The Virgo Consortium: The Millennium Simulation project home page, http://www.mpa-garching.mpg.de/galform/virgo/millennium (2013)
28. L. Perivolaropoulos, LCDM: triumphs puzzles and remedies, in *Proceedings of the Lorenz Center Workshop*, Leiden, Netherlands (2010), arXiv:1104.0539
29. P. Salucci, The mass distribution in spiral galaxies, in *Proceedings of the 244th International Astronomical Union Symposium*, Cardiff, UK (2008), arXiv:0707.4370
30. C. Trachternach et al., Dynamical centers and noncircular motions in THINGS galaxies: implications for dark matter halos. Astron. J. **136**, 2720–2760 (2008), arXiv:0810.2116
31. J.S. Bullock, Notes on the missing satellites problem, in *Proceedings of the XX Canary Islands Winter School of Astrophysics*, Spain (2008), arXiv:1009.4505
32. J.D. Simon, M. Geha, The kinematics of the ultra-faint milky way satellites: solving the missing satellite problem. Astrophys. J. **670**, 313–331 (2007), arXiv:0706.0516

33. S.S. McGaugh, The baryonic Tully-Fisher relation of gas-rich galaxies as a test of LCDM and MOND. Astrophys. J. **143**, 40 (2012), arXiv:1107.2934
34. M. Milgrom, A modification of the Newtonian dynamics as a possible alternative to the hidden mass hypothesis. Astrophys. J. **170**, 365–370 (1983)
35. J.D. Bekenstein, Relativistic gravitation theory for the modified Newtonian dynamics paradigm, Phys. Rev. D **70**, 083509. Erratum-ibid. D **71**, 069901 (2005), arXiv:astro-ph/0403694
36. K. Gödel, An example of a new type of cosmological solutions of Einstein's field equations of gravitation. Rev. Mod. Phys. **21**, 447–450 (1949)
37. F. Hoyle, G. Burbidge, J.V. Narlikar, The basic theory underlying the quasi-steady-state cosmology. Proc. R. Soc. Lond. A **21**, 191–212 (1995)
38. A. Blanchard et al., An alternative to the cosmological concordance model. A&A **412**, 35–44 (2003), arXiv:astro-ph/0304237
39. J.L. Feng, Dark matter candidates from particle physics and methods of detection. Ann. Rev. A&A **48**, 495–545 (2010), arXiv:1003.0904
40. P. Hut, Limits on masses and number of neutral weakly interacting particles. Phys. Lett. B **69**, 85–88 (1977)
41. G. Gelmini, P. Gondolo, DM production mechanisms, in *Particle Dark Matter: Observations, Models and Searches*, ed by G. Bertone, 1st edn. (Cambridge University Press, Cambridge, 2010), pp. 121–141. ISBN: 978-0521763684
42. P. Fayet, S. Ferrara, Supersymmetry. Phys. Rept. **32**, 249–334 (1977)
43. H.P. Nilles, Supersymmetry, supergravity and particle physics. Phys. Rept. **110**, 1–162 (1984)
44. E. Witten, Mass hierarchies in supersymmetric theories. Phys. Lett. B **105**, 267–271 (1981)
45. J. Ellis, S. Kelley, D.V. Nanopoulos, Constraints from gauge coupling unification on the scale of supersymmetry breaking. Phys. Lett. B **287**, 95–100 (1992), arXiv:hep-ph/9206203
46. J.R. Ellis et al., Supersymmetric relics from the Big Bang. Nucl. Phys. B **238**, 453–476 (1984)
47. P. Fayet, Supersymmetry and weak, electromagnetic and strong interactions. Phys. Lett. B **64**, 159–162 (1976)
48. J. Wess, J. Bagger, *Supersymmetry and Supergravity*, 2nd revised edn. (Princeton University Press, New Jersey, 1992). ISBN: 978-0691025308
49. ATLAS Collaboration: Search for squarks and gluinos using final states with jets and missing transverse momentum with the ATLAS detector in $\sqrt{s} = 7$ TeV proton-proton collisions, ATLAS note ATLAS-CONF-2012-033, CERN, Geneve, Switzerland (2012)
50. A.H. Chamseddine, R.L. Arnowitt, P. Nath, Locally supersymmetric grand unification. Phys. Rev. Lett. **59**, 970974 (1982)
51. J. Ellis et al., Exploration of the MSSM with non-universal Higgs masses. Nucl. Phys. B **652**, 259–347 (2003), arXiv:hep-ph/0210205
52. C.F. Berger et al., Supersymmetry without prejudice. JHEP **02**, 23 (2009), arXiv:0812.0980
53. T. Bringmann, L. Bergström, J. Edsjö, New gamma-ray contributions to supersymmetric dark matter annihilation. JHEP **01**, 049 (2008), arXiv:0710.3169
54. CERN European Organization for Nuclear Research: The Large Hadron Collider home page, http://public.web.cern.ch/public/en/LHC/LHC-en (2013)
55. S. Chatrchyan et al., Observation of a new boson at a mass of 125 GeV with the CMS experiment at the LHC. Phys. Lett. B **716**, 30–61 (2012), arXiv:1207.7235
56. R. Aaij et al., First evidence for the decay $B_s^0 \rightarrow \mu^+\mu^-$. Phys. Rev. Lett. **110**, 021801 (2013), arXiv:1211.2674
57. CERN Courier: Work for the LHCs first long shutdown gets under way (2013)
58. T. Kaluza, On the problem of unity in physics, Sitzungsber. Preuss. Akad. Wiss. Berlin (Math. Phys.), pp. 966–972 (1921)
59. O. Klein, Quantum theory and five-dimensional theory of relativity. Z. Phys. **37**, 895–906 (1926)
60. T. Appelquist, H.C. Cheng, B.A. Dobrescu, Bounds on universal extra dimensions. Phys. Rev. D **64**, 035002 (2001), arXiv:hep-ph/0012100
61. J.A.R. Cembranos, A. Dobado, A.L. Maroto, Brane-world dark matter. Phys. Rev. Lett. **90**, 241301 (2003), arXiv:hep-ph/0302041

62. M. Schmaltz, D. Tucker-Smith, Little Higgs review. Ann. Rev. Nucl. Part. Sci. **55**, 229–270 (2005), arXiv:hep-ph/0502182
63. K. Agashe, G. Servant, Warped unification, proton stability, and dark matter. Phys. Rev. Lett. **93**, 231805 (2004), arXiv:hep-ph/0403143
64. T. Asaka, M. Shaposhnikov, The νMSM, dark matter and baryon asymmetry of the Universe. Phys. Lett. B **620**, 17–26 (2005), arXiv:hep-ph/0505013
65. M. Göetz, J. Sommer-Larsen, Galaxy formation: warm dark matter, missing satellites, and the angular momentum problem. Astrophys. Space Sci. **284**, 341–344 (2003), arXiv:astro-ph/0210599
66. A. Dolgov, S. Hansen, Massive sterile neutrinos as warm dark matter. Astropart. Phys. **16**, 339–344 (2002), arXiv:hep-ph/0009083
67. J.L. Feng, A. Rajaraman, F. Takayama, Super WIMP dark matter signals from the early Universe. Phys. Rev. D **68**, 063504 (2003), arXiv:hep-ph/0306024
68. C. Weinheimer, Introduction to gravitational lensing and cosmology, in *Proceedings of the 10th International Workshop on Neutrino Telescope*, Venice, Italy (2003)
69. R.D. Peccei, H.R. Quinn, CP conservation in the presence of pseudoparticles. Phys. Rev. Lett. **38**, 1440–1443 (1977)
70. J.L. Feng, J. Kumar, Dark-matter particles without weak-scale masses or weak interactions. Phys. Rev. Lett. **101**, 231301 (2008), arXiv:0803.4196

Chapter 2
Dark Matter Searches

The last decades have been marked by ever-growing efforts to discover the true nature of dark matter. Numerous experiments have been devised in attempt to catch a glimpse of the elusive dark matter particle, however, to date, no undeniable results can be claimed. Based on the approach, three main detection techniques can be distinguished: production of dark matter in particle accelerators, direct detection through dark matter scattering off ordinary matter, and indirect detection of primary or secondary SM particles produced in dark matter annihilation or decay.

This chapter describes the basic principles behind each of these approaches, as well as their latest experimental results. As implied from the title of this work, indirect searches will be discussed in more detail—from the technique and expected spectral signatures, to suitable targets, existing observatories and the most interesting results.

2.1 Production at Particle Colliders

Possible detection of new physics in particle collider experiments may shed some light on the nature of dark matter. If WIMP (Sect. 1.3.1) is the dark matter particle, it could be created in a collider whose luminosity and center-of-mass energy are sufficiently large. The produced WIMPs would, of course, be invisible, but their presence might be deduced indirectly, by measuring the outcome of the collisions (see, e.g. [1] and references within).

The *missing transverse energy* refers to the energy carried away by a body leaving the detector unseen. It is reconstructed from the momentum conservation law: the momenta of incoming projectiles in the direction orthogonal to the beam is zero, so the final products of the collision must balance their momenta in the transverse plain. When this does not happen, a possible explanation is the production of dark matter particles.[1]

[1]Other explanations include the escape of high energy neutrino or imperfect reconstruction of the momenta (if the missing transverse energy is not too large and significantly different from zero.).

J. Aleksić, *Optimized Dark Matter Searches in Deep Observations of Segue 1 with MAGIC*, Springer Theses, DOI 10.1007/978-3-319-23123-5_2

Different mechanisms for WIMP production at colliders have been proposed. In the particular case of SUSY (Sect. 1.3.1.1), the best detection prospects would arise from the creation of heavier superparticles that in turn decay into quark and gluon jets and pair(s) of neutralinos. Such events should be seen at the LHC detectors ATLAS [2] and CMS [3]; however, failure to detect them so far has put serious strains on the most simplified SUSY models [4]. The latest limits from ATLAS on mSUGRA model [5] allow for the exclusion of a large region of the parameter space (Fig. 2.1). Still, more on validity of SUSY will most likely have to wait until 2015, when LHC will begin operating at its full energy potential of 14 TeV [6].

Collider experiments also offer the possibility of probing the UED models (Sect. 1.3.1.2). The expected signatures in this case are remarkably similar to those of SUSY, and involve jets, leptons and missing energy. In the hypothetical case of the discovery of new physics, SUSY and UED particles would be distinguished by their different spins or by different number of partners for each SM particle (infinite tower of particles for UED versus one partner for SUSY).

Also possible is the direct production of WIMP particle pair accompanied by emission of a photon or a jet from the initial state. Such mono-photon and mono-jet events, respectively, together with the missing transversal energy carried away by the WIMPs, would represent striking signatures of dark matter presence. In lepton colliders, disentanglement of such signals from the background is possible, as the initial state particles have definite energy and may be polarized, which provides useful diagnostics. For the hadron colliders, however, these features are missing, since energies of the gluons and quarks, that actually interact in the collision, are not fixed. Consequently, in LHC, mono-jet and mono-photon signals are highly obscured by the background. Still, limits for such events can be made, and they are directly comparable to the constraints of direct search experiments [7].

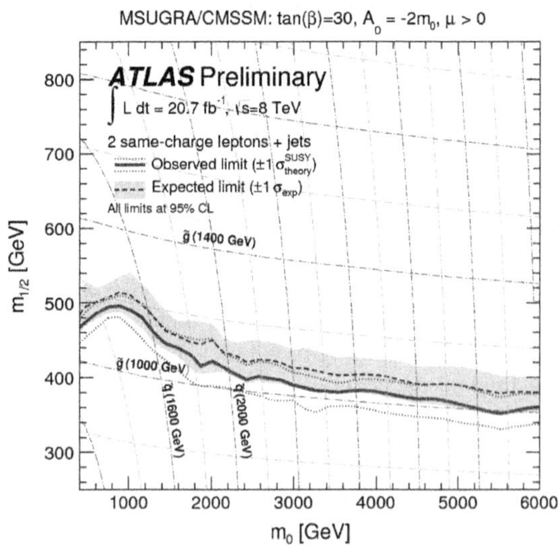

Fig. 2.1 ATLAS exclusion limits over the mSUGRA/cMSSM parameter space after 20.7 fb^{-1} of accumulated data. The *yellow* band around the expected limit shows the $\pm 1\sigma$ uncertainty region, including all statistical and systematic uncertainties except the theoretical uncertainties, on the SUSY cross section. The $\pm 1\sigma$ lines around the observed limit are obtained by changing the SUSY cross section by $\pm 1\sigma$. Taken from [5]. (Color in online)

Production and detection of the dark matter particle in colliders would reveal significant information, like its mass, annihilation and direct detection cross section, as well as the value of its thermal relic density. Nevertheless, such set of characteristics would have to be independently confirmed by direct and indirect detection experiments before identifying the new particle as dark matter.

For more details on collider searches, refer to e.g. [1].

2.2 Direct Detection

If dark matter is made of WIMPs, then the WIMP flux expected on Earth is of the order of $10^5\ \mathrm{cm}^{-2}\ \mathrm{s}^{-1}$ for a particle of $m_\chi = 100$ GeV [8]. This flux is sufficiently large to have a small, but potentially determinable fraction of WIMPs interact with ordinary matter. Direct detection experiments aim to discover dark matter by measuring the nuclear recoils caused by elastic scattering of the WIMPs off baryonic targets. Assuming that the velocity distribution of WIMPs with respect to the Solar System is of order of $100\,\mathrm{km\,s}^{-1}$, the expected recoil energy, transferred from a GeV-mass WIMP to a heavy nucleus, is typically of order of tens of keV. The energy exchanged in these interactions can be deposited in the detector through ionization, scintillation or heat (phonon) production.

All the information about dark matter microscopic properties is codified into the differential elastic scattering cross section, generally separated into a spin-independent and spin-dependent contributions. The spin-independent term comes from scalar and vector couplings to quarks, and its value basically scales as the square number of nucleons. On the other hand, the spin-dependent term comes from axial-vector couplings to quarks, and it is dependent on the nuclear angular momentum. For different dark matter models relation between these two contributions may differ, and although both have to be taken into account, the scalar component dominates for heavy targets, which is the case for most direct detection experiments.

Since the expected elastic cross section is of order of $\sigma \sim 10^{-43}\,\mathrm{cm}^2$, the rate of nuclear interactions is extremely low (less than 1 event per kg per day). That makes the background characterization and control the greatest challenges of direct detection experiments. Better performance is ensured by choosing a large detection target, composed of extremely radio-pure elements; the same philosophy steers the selection of the rest of the detector parts. Furthermore, the target material is often surrounded by a high-density metal shielding, and special care is taken to minimize the electronic noise. In order to suppress the unwanted background originating from cosmic rays (mainly muons), the installations of the experiments are typically located deep under ground.

By the time this work was written, hints of dark matter signal have been reported by experiments like DAMA/LIBRA [9] and CoGeNT [10]; however, neither was conclusive enough on its own and could not be reproduced by other experiments. Moreover, the most stringent limits over the spin-independent interaction cross section, measured by the XENON100 experiment [11] and of order $\sim 10^{-45}\,\mathrm{cm}^2$ [12],

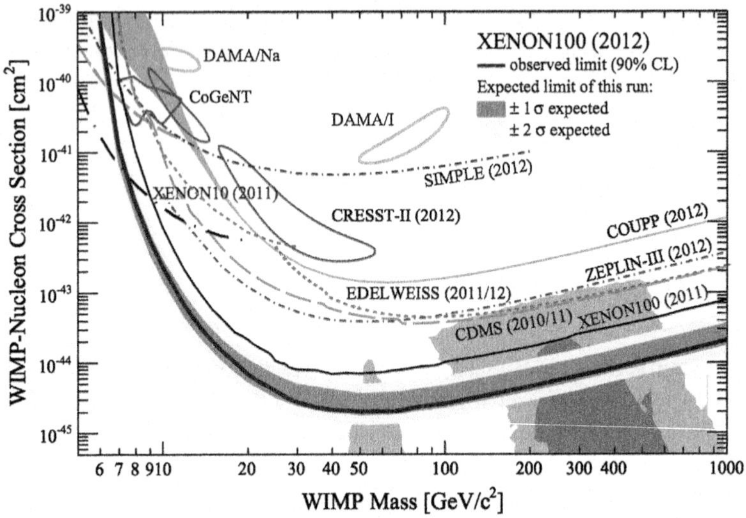

Fig. 2.2 Result on spin-independent WIMP-nucleon scattering from XENON100, from 225 live days. The expected sensitivity is shown by the *green/yellow* band ($1\sigma/2\sigma$) and the resulting exclusion limit (90 % c.l.) in *blue*. For comparison, other experimental limits (90 % c.l.) and detection claims (2σ) are also shown, together with the regions ($1\sigma/2\sigma$) preferred by cMSSM models. Taken from [12]. (Color in online)

already exclude both the DAMA/LIBRA and the CoGeNT favored regions (Fig. 2.2). XENON100 also provides the highest sensitivity for the spin-dependent cross section, of order $\sim 10^{-40}\,\text{cm}^2$ [13].

The future of the direct detection instruments goes along the line of increasing the mass of the target materials above one tone, lowering the ambiental temperature down to few mK, and measuring the signals from ionization, scintillation and heat production within the same detector. Efforts on several of such future experiments, like EURECA [14], DARWIN [15] and XENON1T [16], are already under way.

For more information on direct detection searches, see e.g. [8] and references within.

2.3 Indirect Detection

Indirect searches exploit the possibility that dark matter particles can annihilate or decay, producing SM particles detectable through a variety of modern ground and space-based observatories. The resulting SM products are expected to carry valuable information that could give clues about the properties of dark matter particle. Furthermore, indirect searches are probing the astrophysical distribution of dark matter, which is not possible with direct and collider approaches. The main obstacle to this search method is the (usually) overwhelming abundance of astrophysical background, which makes the disentanglement of SM particles that are of dark matter origin a rather complex task.

The premise of dark matter particle annihilating or decaying is based on the assumption that this particle is not absolutely stable, but *stable on cosmological scales*, i.e. its lifetime is longer than the age of the Universe and its annihilation rate is sufficiently small so that the total dark matter budget is not significantly reduced.

Indirect searches look for signatures of dark matter through the stable final SM products: photons, neutrinos, electrons, protons and their corresponding antiparticles. The expected signal depends on the properties of the dark matter particle, on the resulting final state SM particle, as well as on how and where it was produced. This work focuses on searches for gamma-ray signatures of dark matter annihilation or decay. The remaining of this chapter describes the calculation of the expected photon flux, the suitable targets and types of gamma-ray observatories used for indirect searches. But first, for the sake of completeness, the most relevant detectable products are listed in the following section.

2.3.1 Messengers for Indirect Dark Matter Searches

Photons are particularly interesting products of dark matter annihilation or decay, as they travel in straight lines and are practically unabsorbed in the local Universe. Because they point back to the place of their creation, astrophysical foregrounds can be significantly reduced by looking for signals in regions with high dark matter density. Furthermore, the resulting photon spectrum should bare some characteristic features (Sect. 2.3.2.3), unique and universal for dark matter annihilation or decay, whose detection would represent the 'smoking gun' of indirect searches. For WIMP-type dark matter, emission of photons is expected in the gamma-ray energy range. More details on the gamma-ray-based searches are provided in the following sections.

Neutrinos, like photons, are not deflected by magnetic fields and thus can be traced back to their source of origin. Neutrinos do not couple with electromagnetic sector and their interactions with matter are weak, however, they could potentially be detected in highly transparent well-shielded deep water (ANTARES [17]) or ice (IceCube [18]) detectors.

Neutrinos are expected to be produced in large amounts in dark matter annihilation or decay. If primary products from these processes are heavy leptons, their consequent decay into lighter ones will be accompanied by neutrino emission. If the primary products are gauge bosons, neutrinos are also produced in their decay into lepton (for $W^{\pm}$, Z) and quark pairs (for Z). In addition, if Z boson is among the primaries, it can decay directly into a pair of neutrinos. Direct annihilation into a neutrino pair is possible as well.

Neutrinos can also be produced by dark matter that gets captured by deep gravitational wells, such as the Sun, and that annihilates at significant rate if gathered in great concentrations. Neutrinos can escape compact objects, and a detection of a neutrino excess from the direction of the Sun could indicate dark matter origin [19].

The same reasoning can be applied to dark matter captured by the Earth, but the detection prospects are much weaker than in the case of the Sun.

The currently best limits on dark matter annihilation cross section from neutrino searches come from IceCube observation of the Galactic Center [20]: for $m_\chi \sim$ 200 GeV, $\langle\sigma_{\text{ann}}v\rangle$ for direct annihilation into neutrinos is $\sim 10^{-23}$ cm^3 s^{-1}, while the lower limit on lifetime of the dark matter particle in the decay scenario is $\tau_\chi \sim 10^{22}$ s.

Charged Cosmic Rays diffuse through the galactic magnetic field from their production site to the Solar System, so, unlike photons and neutrinos, they can not be traced back to the place of their origin. It therefore makes sense to search for dark matter signal as an anomalous component in isotropic cosmic ray spectrum.

Given that dark matter annihilation or decay results in the creation of the same amounts of matter and antimatter, the latter products are especially attractive from the point of indirect searches, since for them the astrophysical background is much lower. In this sense, distributions of positrons and antiprotons are very promising places to look for deviations from conventional flux expected from astrophysical processes. In the last years, there have been a number of reports on unusual features in the electron-positron spectrum at high energies. The PAMELA experiment (Payload for Antimatter Matter Exploration and Light-nuclei Astrophysics, [21]) found an interesting rise in the positron fraction ($e^+/(e^+ + e^-)$) at energies up to 100 GeV [22], a behavior in contradiction to the expected decline predicted by traditional models of cosmic-ray propagation [23] (Fig. 2.3). This result was corroborated by measurements by Fermi-LAT (Sect. 2.3.4.1) for energies up to 200 GeV [24]. The latest news on this subject come from the high-precision results of AMS-02 (Alpha Magnetic Spectrometer-02, [25]) that extend up to 350 GeV [26]: these measurements confirm the rise for energies up to ~250 GeV, above which there is a hint of spectrum flattening (Fig. 2.3). There are numerous proposed theories involving dark matter [27, 28] that can justify the observed excess, however, they are not fully supported by the experimental measurements (for instance, the positron excess should be accompanied by photon excesses at other wavelengths, which is not the case). On the other hand, a more conventional explanation, with particles being accelerated by the nearby pulsars [29], is much more plausible. Another stable product from dark matter annihilation or decay are the antiprotons. Antiprotons may be created from decay of primary products; however, current measurements of the antiproton flux show no deviation from the predictions for local astrophysical sources [30]. Antideuterium is other possible messenger: if of dark matter origin, its spectrum should be much flatter than the standard astrophysical component.

For more detailed description on indirect searches with cosmic antimatter, refer to e.g. [31] and references within.

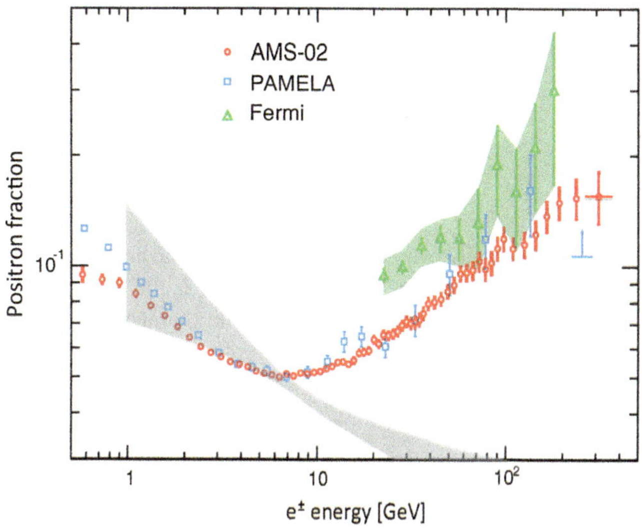

Fig. 2.3 The positron fraction in high-energy cosmic rays. The measurement from the AMS extends over a wider energy range and has much lower uncertainty than the earlier measurements from the PAMELA [22] and Fermi-LAT satellites [24]. The AMS measurement confirms an excess in the high-energy positron fraction, above what is expected from positrons produced in cosmic-ray interactions. (The *grey* band indicates the expected range in the positron fraction, which is based on calculations in [23]). Taken from [26]. (Color in online)

2.3.2 Photon Flux from Dark Matter

As already mentioned, dark matter annihilation or decay into SM particles is expected to produce gamma-ray signal with some distinctive observational features that can be used to obtain information about both the dark matter nature and its spatial distribution. Detection of such signatures would aid the disentanglement of dark matter signal from the astrophysical backgrounds, and potentially allow an unambiguous identification of the dark matter particle. This section gives a brief introduction on the calculation of the expected photon flux on Earth, followed by modeling of dark matter distribution and details on the characteristic spectral shapes resulting from different annihilation/decay final states.

Dark Matter Annihilation The most general form of the expected differential dark matter photon flux is given as a product of two terms:

$$\frac{d\Phi(\Delta\Omega)}{dE} = \frac{d\Phi^{\mathrm{PP}}}{dE} \times J(\Delta\Omega). \tag{2.1}$$

The particle physics term, $d\Phi^{\mathrm{PP}}/dE$, solely depends on the chosen dark matter model—it is completely determined for the given theoretical framework and its value is the same for all sources. The astrophysical term, $J(\Delta\Omega)$, on the other hand, depends on the observed source (its distance and geometry), the dark matter distribution at the source region and the properties of the instrument.

In the case of annihilating dark matter, the first term takes the form:

$$\frac{d\Phi^{PP}}{dE} = \frac{1}{4\pi} \frac{\langle \sigma_{ann} v \rangle}{2m_\chi^2} \frac{dN}{dE}, \tag{2.2}$$

where $\langle \sigma_{ann} v \rangle$ is the thermally averaged product of the total annihilation cross section and the velocity of the dark matter particles; dN/dE is the differential gamma-ray yield per annihilation, summed over all the n possible channels that produce photons, where each channel has its particular branching ratio Br:

$$\frac{dN}{dE} = \sum_{i=1}^{n} \mathrm{Br}_i \frac{dN_i}{dE}. \tag{2.3}$$

The spectral information (i.e. the spectral shape) is described by dN/dE contribution.

As for the astrophysical factor J_{ann}, it is given as the integral over the squared dark matter density profile ρ over the line of sight l and the solid angle $\Delta\Omega$:

$$J_{ann}(\Delta\Omega) = \int_{\Delta\Omega} \int_{los} \rho^2(l, \Omega) dl d\Omega. \tag{2.4}$$

Finally, the integral flux from dark matter annihilation (above a certain energy E_0) reads:

$$\Phi(> E_0, \Delta\Omega) = \frac{1}{4\pi} \frac{\langle \sigma_{ann} v \rangle}{2m_\chi^2} \int_{E_0}^{m_\chi} \frac{dN}{dE} dE \int_{\Delta\Omega} \int_{los} \rho^2(l, \Omega) dl d\Omega. \tag{2.5}$$

Dark Matter Decay Assuming that dark matter particles are not completely stable (that is, their stability is valid on cosmological scale), a small fraction of them could be decaying into detectable final states. Like in the case of annihilation, the expected gamma-ray flux has the general form given by the Eq. (2.1); however, the particle physics and the astrophysical term are different. The particle physics term depends on the lifetime of the particle τ_χ:

$$\frac{d\Phi^{PP}}{dE} = \frac{1}{4\pi} \frac{1}{m_\chi \tau_\chi} \frac{dN}{dE}, \tag{2.6}$$

while the astrophysical term scales linearly with the dark matter density:

$$J_{dec}(\Delta\Omega) = \int_{\Delta\Omega} \int_{los} \rho(l, \Omega) dl d\Omega. \tag{2.7}$$

The total flux within a solid angle $\Delta\Omega$ above a given energy is then calculated in a manner similar to Eq. (2.5).

2.3.2.1 Dark Matter Density Profile

In the above described calculations of the photon flux, the greatest uncertainty arises from the poorly determined dark matter density distribution. This distribution is not directly observable and can only be constrained by N-body simulations and stellar and gas kinematics.

The N-body simulations recreating the hierarchical formation of CDM halos through gravitational interactions have shown that the spherically-averaged (smoothed) dark matter halo distribution is well described by a universal profile. Navarro, Frank and White (NFW, [33]) have made a fit to this profile, applicable over 20 decades in mass range:

$$\rho_{\mathrm{NFW}}(r) = \rho_s \left(\frac{r}{r_s}\right)^{-1} \left(1 + \frac{r}{r_s}\right)^{-2}, \tag{2.8}$$

where r_s is the scale radius and ρ_s is the characteristic density. For $r \ll r_s$, the NFW profile has a central power-law cusp with $\rho_{\mathrm{NFW}} \propto r^{-1}$, and at large radii ($r \gg r_s$) it declines as $\rho_{\mathrm{NFW}} \propto r^{-3}$ (Fig. 2.4). The J factor (especially for dark matter annihilation) is primarily sensitive to the behavior of the density profile in the cusp region; however, the current CDM simulations have limited capability to model the dark matter distribution on such scales. The recent sample of Aquarius simulations [34], which resolve down to less than 1 % of the halo viral radius for

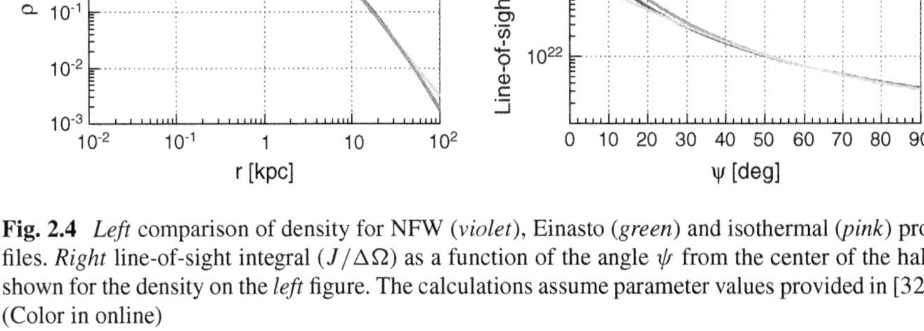

Fig. 2.4 *Left* comparison of density for NFW (*violet*), Einasto (*green*) and isothermal (*pink*) profiles. *Right* line-of-sight integral ($J/\Delta\Omega$) as a function of the angle ψ from the center of the halo shown for the density on the *left* figure. The calculations assume parameter values provided in [32]. (Color in online)

$\sim 10^{12}$ M$_\odot$ halos, shows that the asymptotic slope, predicted by the NFW, is not yet achieved, and that the profile becomes more and more shallow towards the center of the halo. These results favor the Einasto profile [35] as a better fit:

$$\rho_{\rm Ein}(r) = \rho_s \exp\left[-\frac{2}{\alpha}\left(\left(\frac{r}{r_s}\right)^{\alpha} - 1\right)\right], \tag{2.9}$$

where $\alpha \simeq 0.17$ for Milky Way-mass halos. The slope of the Einasto fit asymptotically approaches zero toward the center, producing a finite density at $r = 0$ (unlike the divergent central density of NFW). Still, the NFW and Einasto profiles are similar over the intermediate scales ($r \sim r_s$, Fig. 2.4).

However, the predicted central cusp is not observed in many galaxies. In particular, observations of low surface brightness galaxies indicate inner slopes that are significantly shallower than expected (based on N-body simulations). The isothermal profile [36], characterized by the constant velocity dispersion, has, in contrast to NFW and Einasto profiles, a central cored density distribution (Fig. 2.4):

$$\rho_{\rm iso} = \rho_s\left[1 + \left(\frac{r}{r_s}\right)^2\right]^{-1}. \tag{2.10}$$

This discrepancy between the simulations and observations, i.e. cusped versus cored central densities, is an issue that is yet to be understood [37]. The density profile of a dark matter halo is determined by a number of astrophysical processes, such as the initial gravitational relaxation of dark matter, its interaction with baryons, and potentially the weak interactions of WIMPs with themselves and other particles. The inclusion of all of these factors in the N-body simulations with sufficiently high resolution is a notoriously difficult task, but a necessary one needed to fully resolve the core/cusp problem.

The choice of the density profile has direct implications on the expected photon flux, in particular in the case of dark matter annihilation (Eq. (2.5)): as J factor is proportional to the density squared, cored central distributions will yield lower fluxes than the cusped ones. This effect is less pronounced for the decaying dark matter, as in that case the flux scales linearly with ρ.

Substructures Another factor that may influence the expected photon flux is the presence of substructures: a generic prediction of CDM simulations is that dark matter distribution is not smooth, but rich in density fluctuations with spatial scales smaller than r_s (Fig. 2.5). These substructures are predicted to be gravitationally isolated; their abundance at $z = 0$ depends on the fraction that survives tidal disruptions during the hierarchical mergers and accretion processes [38].

The presence of substructures may significantly enhance the dark matter flux (especially in the annihilation case), compared to the expectations for smooth halos. This contribution is quantified by a boost factor B, such that $J = J_s(1 + B)$, where J_s is the line-of-sight integral over the smooth halo distribution. N-body simulations can set a lower limit on the B value by summing the annihilation luminosity from

Fig. 2.5 Projected dark matter density in high resolution Milky Way-sized halo at $z = 0$ from Aquarius simulation. The halo is resolved with $\sim$1.5 billion particles of 1712 $M_{\odot}$ each. It contains $\sim$300000 resolved substructures, the largest of which are visible as bright spots on the image. Taken from [38]

numerically resolved substructures; high resolution simulations of Milky Way-mass halos have found a factor $B \sim 2$ [39]. The unresolved substructures could contribute to the value of B more significantly: depending on the halo mass, the boost factor may be in the range $B \simeq 10^2 - 10^3$ for galaxy clusters, $B \sim 10^2$ for Milky Way-mass galaxies and only a factor of few for dSphs (see, e.g. [40] and references within). These estimates, however, heavily rely on the extrapolation of the simulated halo mass spectrum, and thus are subjected to high uncertainties and should be revised once the simulations reach better mass resolution.

2.3.2.2 Annihilation Cross Section and Decay Time

Next, in order to calculate the gamma-ray flux from dark matter annihilation or decay into SM particles, it is necessary to evaluate the $\Phi^{\mathrm{PP}}(E)$ contribution. First, the dark matter particle has to be qualified as scalar (of spin 0) or fermion (of spin 1/2); in the latter case, it is further identified as either Dirac (has antiparticle) or Majorana (is its own antiparticle). From there, the interaction operators can be constructed between the dark matter and SM particles (for a review, see e.g. [41]), which then leads to calculation of the annihilation cross section or decay time.

The natural value of $\langle \sigma_{\mathrm{ann}} v \rangle \simeq 3 \times 10^{-26}$ cm^3 s^{-1} is determined by the requirement to produce the observed dark matter relic abundance during the dark matter freeze-out process in the early Universe [42]. For WIMPs that annihilate at the present time, predominantly through the s-wave processes, $\langle\sigma_{\rm ann}v\rangle$ is velocity independent and has the natural value. This is the case for dark matter annihilation with pseudoscalar and vector interactions. On the other hand, if the cross section depends on the velocity, $\langle\sigma_{\rm ann}v\rangle$ would be lower today than in the early Universe, as the typical velocity in a halo at present is $v/c \simeq 10^{-3}$, opposed to $v/c \simeq 0.1$ at the time of WIMP freeze-out. Some exceptions exist, though: the Sommerfeld enhancement, a non-relativistic quantum mechanical effect, can boost the $\langle\sigma_{\rm ann}v\rangle$ value, by a factor $\sim$10–100, due to the resonance annihilation for certain, heavy dark matter masses, assuming that the WIMP velocity is small enough [43].

Regarding the decay time, its value depends on the considered model and dark matter particle properties, but, as already mentioned, it has to be longer that the current age of the Universe ($(4.35 \pm 0.01) \times 10^{17}$ s, [44]). For example, in a SUSY extension of the SM with small R-parity violation and gravitino as LSP, decay time into photon and neutrino is calculated to be of order $\tau_{\psi_{3/2}\rightarrow\gamma\nu} \simeq 3.8\times10^{27}$ s [45].

2.3.2.3 The Photon Spectrum

The spectral distribution of photons emitted in dark matter annihilation or decay is one the decisive factors for detectability of a given model. The characteristic spectral features, that can not be imitated by the conventional astrophysical mechanisms of gamma-ray production, can be used to distinguish the signal of dark matter origin from the astrophysical background, while at the same time revealing information on the dark matter nature itself. The spectral distribution dN/dE is fixed for a chosen model and universal for all dark matter emitters. As given by Eq. (2.3), dN/dE is the sum over different final-state contributions; depending on the branching ratio of each of these channels, different features will be more or less pronounced in the spectrum. Some of the main final-state contributions to dN/dE from dark matter annihilation/decay are listed bellow.

Secondary Photons The main photon production channel is through the decay of neutral pions, created in the hadronization of fermion and gauge boson final states. The resulting, *secondary* photons, show a continuous and relatively featureless spectrum with a rather soft cutoff at the kinematical limit $E = m_\chi$ (Fig. 2.6). Spectral distributions are very similar for almost all channels and depend very weakly on m_χ. The number of photons produced this way peaks for energies approximately an order of magnitude below m_χ. Still, a convincing claim of dark matter detection based exclusively on this signal, which would show up as a broad bump-like excess over the (often) poorly understood astrophysical background, may be considered rather challenging.

Another contribution to this continuous distribution comes from the final state radiation (FSR)—emission of an additional photon (from the external legs) whenever dark matter annihilates into charged particles (Fig. 2.7). The FSR spectral distribution is broad, model-independent, and unlike the emission from hadronization, it peaks

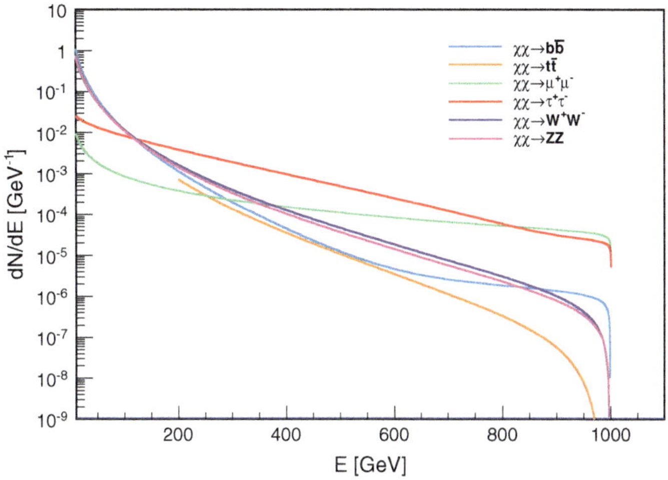

Fig. 2.6 Gamma-ray yield from WIMP annihilation into different channels. The assumed dark matter particle mass is $m_\chi = 1$ TeV. When applicable, the FSR is included in the spectrum. Spectral distributions are obtained from the fits provided in [46]

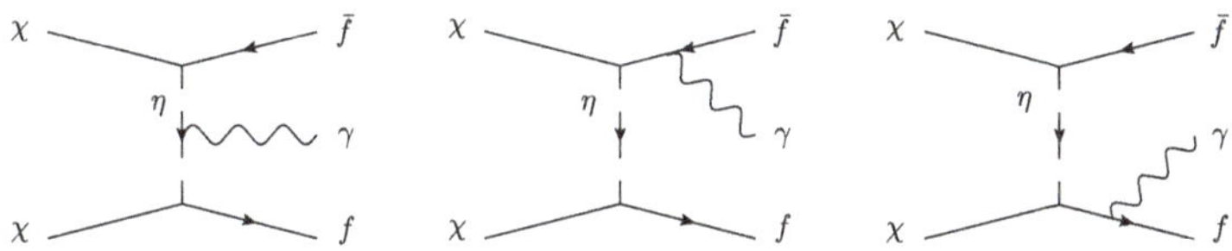

Fig. 2.7 Feynman diagrams for the VIB (*left*) and FSR processes (*centre* and *right*). Taken from [47]

near m_χ and ends with a sharp cutoff (Fig. 2.8). Although the production of FSR photons is suppressed by a factor α ($\approx$1/137) relative to the emission from secondary photons [48], this contribution is the dominant radiation for some channels, like $\chi\chi \rightarrow \mu^+\mu^-$.

Gamma-ray Line Direct production of gamma-rays through dark matter annihilation (or decay) is a highly suppressed process ($\mathcal{O}(1/\alpha^2)$), as dark matter particles do not couple to photons directly. The branching ratio for such one-loop interactions is in general negligible compared to annihilation to fermions or bosons. Nevertheless, if such process was to occur, the result would be a sharp, monochromatic line-like feature in the photon spectrum—a feature whose detection would represent the *smoking gun* for dark matter indirect searches.

Gamma-ray lines are created in direct WIMP annihilation into two photons, $\chi\chi \rightarrow \gamma\gamma$, or into photon and a boson, $\chi\chi \rightarrow \gamma(Z, h)$; such lines would be located at energies $E_\gamma = m_\chi$ and $E_\gamma = m_\chi(1 - m^2_{(Z,h)}/4m^2_\chi)$, respectively. The $\langle\sigma_{\rm ann}v\rangle$ needed in these cases, however, is orders of magnitude smaller than for the tree-loop channels contributing to the continuum component of the spectrum: $\langle\sigma_{\rm ann}v\rangle \sim 10^{-29}$ cm^3 s^{-1}. As for the dark matter decay, line production is also a possibility

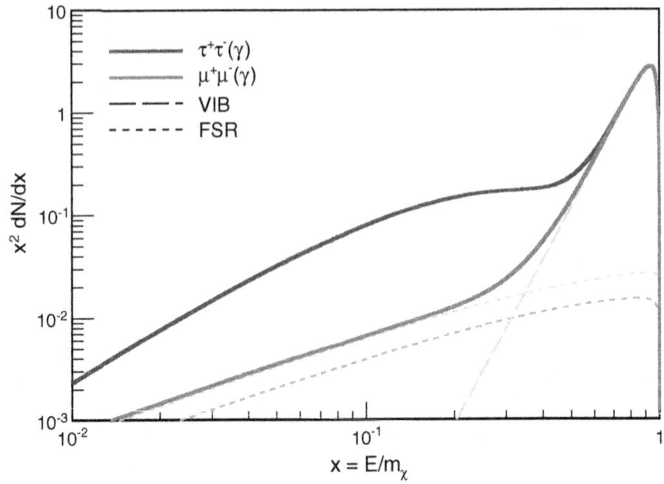

Fig. 2.8 Gamma-ray spectrum for different final state fermions, assuming $m_\chi = 100$ GeV and a mass-splitting of $\mu = 1.1$. *Solid lines* show the full contribution from three-body final states, including the VIB photons close to $x = 1$; *short-dashed lines* depict the FSR contribution, while the *long-dashed* ones represent the VIB emission

($\chi \to \gamma\nu$, with $E_\gamma = m_\chi/2$). It is needed, however, for dark matter particle to have a lifetime τ_χ of 10^{27} s or longer [45].

Virtual Internal Bremsstrahlung The virtual internal bremsstrahlung (VIB) process occurs when the helicity suppression is lifted from the s-wave contribution to $\sigma_{\rm ann}$ [47, 49]. If dark matter is a Majorana (or scalar) particle annihilating into fermion and antifermion ($\chi\chi \to f\bar{f}$), the $\langle\sigma_{\rm ann}v\rangle$ is quite small, as the s-wave contribution is suppressed by the m_f^2/m_χ^2 term (m_f is the mass of the daughter fermion), while the p-wave contribution is suppressed by the v^2 of the dark matter particles today ($v \sim 10^{-3}$). However, emission of a vector boson ($\chi\chi \to f\bar{f}V$) lifts the suppression of the s-wave contribution (Fig. 2.7). The resulting 3-body $\sigma_{\rm ann}$ is, unlike FSR, model-dependent: its value is determined by m_χ and the mass splitting parameter μ between the dark matter particle and the t-channel mediator η ($\mu \equiv m_\eta^2/m_\chi^2$). For small values of μ, the VIB contribution becomes very significant, and $(\sigma v)_{\chi\chi\to f\bar{f}\gamma}$ considerably larger than $(\sigma v)_{\chi\chi\to f\bar{f}}$. Furthermore, the VIB spectral shape has a characteristic bump-like feature close to the cutoff ($E \simeq m_\chi$), that, depending on the model, can be quite pronounced (Fig. 2.8).

Gamma-ray Boxes If dark matter annihilates into a pair of intermediate, neutral scalar particles ϕ, that in turn decay into a pair of photons (Fig. 2.9), the result from this 1-step cascade is a box-shaped spectrum [50]. The width of such feature is completely determined by the mass of the scalar m_ϕ and the dark matter particle m_χ: $\Delta E = \sqrt{m_\chi^2 - m_\phi^2}$—the smaller the mass difference, the more line-like is the resulting spectrum. On the other hand, when $m_\chi/m_\phi \to 0$, the box-like feature becomes wider in energy and dimmer in amplitude; still, the spectral plateau is of

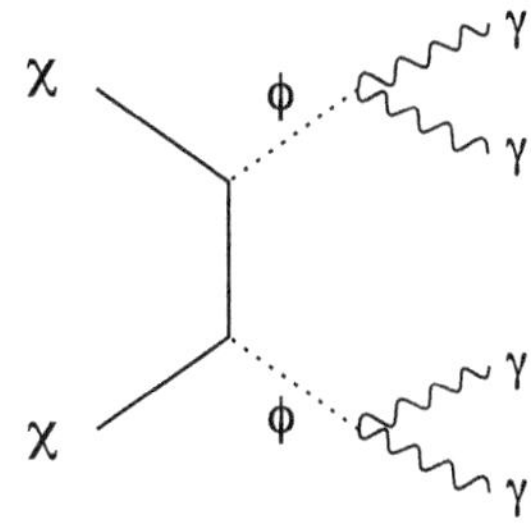

Fig. 2.9 Diagram for annihilation into two intermediate scalars ϕ, that in turn decay into 2 photons each

non-negligible intensity, it can extend to high energies and its flatness may distinguish it from the exponentially-falling astrophysical backgrounds.

As for the case of dark matter decay, the same considerations apply, but instead of four, two photons are produced: $\chi \rightarrow \phi \rightarrow \gamma\gamma$.

2.3.3 Review of the Observational Targets

When choosing a target source for indirect dark matter searches, the selection criteria should go in the direction of maximizing the value of the astrophysical factor J. Therefore, following the prescription from Eqs. (2.4) and (2.7), suitable target should be a region with high dark matter density, while at the same time its distance from the observer is as (relatively) small as possible. The M/L of the system and the possible background must be regarded as well, since the large baryonic content may cause major drawbacks for dark matter searches: baryonic matter may disrupt the dark matter profile through the dynamical friction, smoothing the central high dark matter density, and thus reducing the expected flux. Furthermore, baryons may act as strong backgrounds to the dark matter signal, as they can produce photons via conventional astrophysical processes in a far more abundant fashion than the dark matter annihilation or decay can.

In practice, a compromise between these selection criteria must be reached. The best dark matter candidate sources proposed so far include the Galactic Center and the Galactic Halo, the dark halo substructures, dSph galaxies and galaxy clusters. This section briefly reviews their strengths and weaknesses as potential targets for indirect dark matter searches.

The Galactic Center and Galactic Halo The Galactic Center is the closest known ($\sim$8.5 pc) region highly dominated by dark matter. Theoretical arguments and numerical simulations predict a central dark matter cusp, that would strongly enhance the annihilation signal. However, the Galactic Center is a densely populated region, with large background from conventional sources present at all wavelengths. Furthermore, the great baryonic content in the innermost parts of the Galaxy and the presence of a black hole in its center inevitably lead to the modification of the dark matter profile, making the characterization of the density distribution highly uncertain [51].

Observations of the Galactic Center at the very high energy (VHE) range, where WIMP signatures are expected, have already been carried out by several Cherenkov telescopes, (see, e.g. results from H.E.S.S. (High Energy Stereoscopic System, [52]) and MAGIC (Major Atmospheric Cherenkov Imaging Gamma-ray telescopes, [53])). A non-variable signal was confirmed, with a hard power law spectrum extending up to 20 TeV, which disfavors its dark matter origin. In addition, the spatial extension of the source does not agree with the dark matter profile, and the detected flux is several orders of magnitude above the predictions for the pure dark matter emission. A more likely origin of this signal is the radiation from conventional counterparts in the vicinity of the Galactic Center (e.g. the super-massive black hole Sag A*, the supernova remnant Sag A East and the pulsar with nebulae G359.95-0.04), which completely hide the potential dark matter signatures.

A way to overcome the background contamination is by directing the search to regions which are outside the Galactic plane (and hence unpolluted by astrophysical sources), but, for the annihilation case, still close enough to the Galactic Center to profit from high dark matter density. Such approach has been applied in the observations by H.E.S.S. [54], and it provided the currently most constraining limits on the $\langle\sigma_{\rm ann}v\rangle$ value for considered final state channel in the given energy range (Fig. 2.10) [55]. On the other hand, if studying the dark matter decay, search should be more effective if carried out in direction of poles, i.e. at higher latitudes [45].

In the past year a hint of a monochromatic gamma-ray signal at around 130 GeV was claimed in the Fermi-LAT data of the Galactic Center region (Fig. 2.11, [47, 56]).

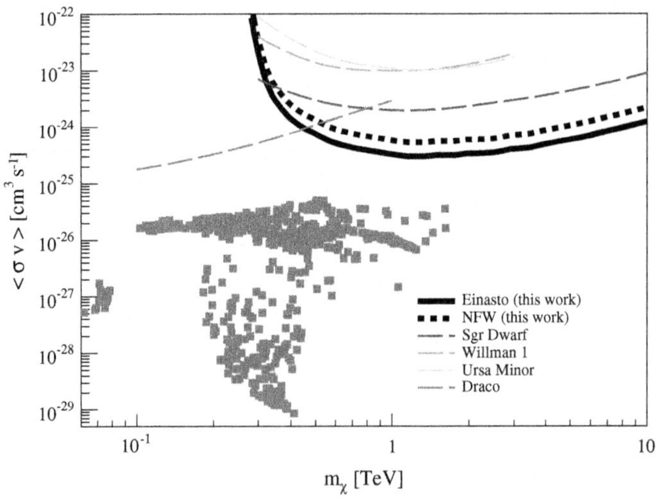

Fig. 2.10 Upper limits at the 95 % c.l. over $\langle\sigma_{\rm ann}v\rangle$ as a function of m_χ, from H.E.S.S. observations of the Galactic Halo, considering a fixed, generic dark matter model for annihilation into quark-antiquark pairs and Einasto and NFW density profiles. The best sensitivity is achieved at $m_\chi \sim$ 1 TeV. For comparison, limits derived from observations of dSph galaxies are also shown. *Green* points represent different simulated mSUGRA models. Taken from [55]. (Color in online)

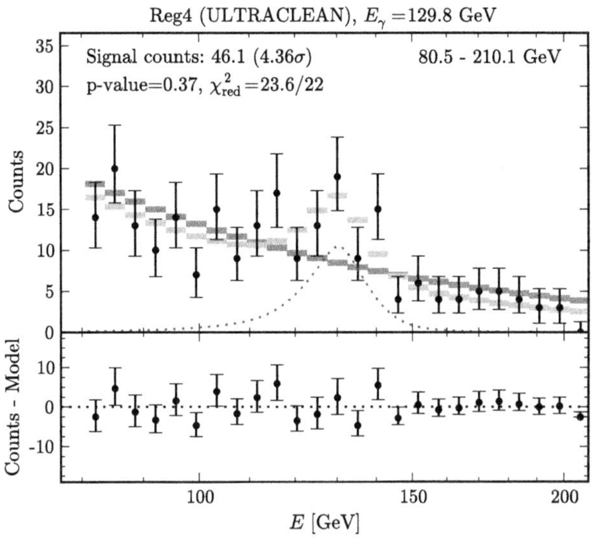

Fig. 2.11 Tentative hint of a line at 130 GeV from 43 months of the Fermi-LAT observations of the Galactic Center region. Taken from [56]

In numerous interpretations that followed, the signal has been attributed to, among others: the VIB contribution, assuming $m_\chi = 149 \pm 4^{+8}_{-15}$ GeV [47]; to $\gamma\gamma$ line, for $m_\chi = 129.8 \pm 2.4^{+7}_{-13}$ GeV [56], and to gamma-ray boxes from axion particles, for $m_\chi = 250$ GeV and $m_\phi = 249.75$ GeV [57]. Decaying dark matter explanations have been postulated as well [58]. The non-dark matter origin of the observed excess is also considered [59]. However, the latest results from the Fermi Collaboration cannot confirm the presence of the signal in question [60]. New information is pending the Pass 8 analysis.

Dark Matter Subhalos According to the N-body simulations, the evolution of dark matter distribution in the Universe is marked by hierarchical clustering, that results in formation of dark matter clumps within halos at all scales (Sect. 2.3.2.1). The larger clumps might gather enough mass to attract baryons and commence star formation, while the smaller ones do not have enough gravitational pull and therefore remain completely dark. As a result, dark matter halos are not expected to be smooth, but rich in inner substructures—the so-called dark matter subhalos (Fig. 2.5). While invisible in the context of conventional emission mechanisms, these subhalos may shine in the energy window where dark matter signal is expected. In the framework of a GeV-mass WIMP, these objects could be gamma-ray emitters. Complete lack of astrophysical background makes the subhalos excellent targets for indirect searches; furthermore, although small, a fraction of them could be relatively nearby. The drawback are their unknown locations; however, subhalos may appear in gamma-ray sky surveys.

Indeed, Fermi-LAT (Sect. 2.3.4.1) has so far detected hundreds of so-called *unassociated Fermi objects* (UFOs)—sources that radiate at very-high energies but are without known counterpart at other wavelengths. Some of these UFOs are potential candidates for dark matter subhalos—if their emission is not variable, if their spectra is hard and power-law like (as the dark matter features are predicted at energies above the Fermi-LAT range). Furthermore, suitable observational targets are outside the Galactic plane, as the abundance of astrophysical sources there not only adds large background but also makes potential identification of an object more complicated.

Complementary observations of several subhalo-candidate UFOs have been performed by Cherenkov Telescopes (see, e.g. [61]) however, no detection has been reported so far.

Dwarf Spheroidal Galaxies These galaxies are the most dark matter-dominated objects in the Universe known so far [63]. With M/L of order of 100 or even 1000 $M_{\odot}/L_{\odot}$, their relative closeness (up to ~250 kpc distance from Earth), and almost non-existent gamma-ray background (as there is no gas and no recent star formation), these systems are considered to be excellent targets for indirect dark matter searches. Furthermore, as baryons are the subdominant component in a dSph, and are not expected to significantly alter the dark matter density profile, studies of their stellar kinematics can be used to set relatively robust constraints on the values of the astrophysical factor J [64].

Currently the most constraining limits from dSphs are set by the Fermi-LAT joint analysis of observations of 10 dwarf galaxies [62], for m_{χ} up to 1 TeV (Fig. 2.12). As

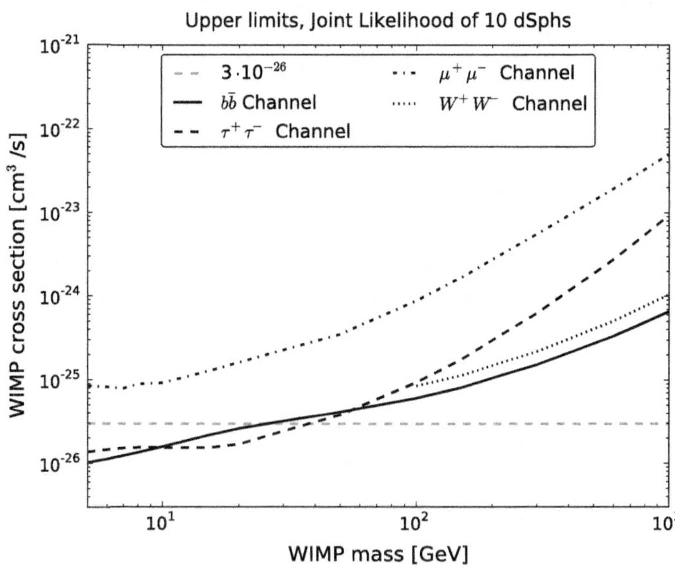

Fig. 2.12 Upper limits on $\langle\sigma_{\mathrm{ann}} v\rangle$ as a function of m_{χ}, for several different final state channels (each with Br = 100 %), from the Fermi-LAT joint analysis of observations of 10 dSph galaxies. Taken from [62]

this work presents the results from indirect search in a dSph galaxy, more complete motivation for these objects as good dark matter targets, as well as other competitive results are presented in Chap. 5.

Galaxy Clusters Galaxy clusters are the largest known gravitationally bound systems, with radii of several Mpc and masses of $\sim(10^{14}-10^{15})$ $M_{\odot}$. In the hierarchical formation of large scale structures (Sect. 1.2.2), these objects represent the top stage as they were the last to form. Galaxy clusters are among the most dark matter dominated objects—more than 80 % of their total mass is dark, while the remaining percentage is divided among galaxies (~5 %) and gas (~15 %). High dark matter content makes them attractive targets for indirect searches [65], with significant astrophysical factor values that can be further increased from the presence of substructures (Sect. 2.3.2.1)—the boost factor can be of order of 100–1000 [66]. However, the role of galaxy clusters as good dark matter targets is weakened by the huge background of astrophysical origin, dominantly from the active galactic nuclei galaxy population, as well as by the secondary component of cosmic ray-induced radiation.

So far, gamma-ray observations of galaxy clusters Virgo, Coma, Perseus and Fornax have not returned any positive signal of dark matter emission [67, 68]. Campaigns at other wavelengths have not claimed any dark matter hints either [69].

2.3.4 Indirect Search with Gamma-Ray Experiments

Annihilation and decay of dark matter particles, whose mass is in the tens of GeV – few TeV range, would result in a production of photons of energies in the high (30 MeV–100 GeV) and very high (100 GeV–30 TeV) subranges of the gamma-ray domain. Seen how the Earth's atmosphere is opaque for such radiation, gamma-rays can not be directly measured from the ground. Instead, detectors have to be placed outside the atmosphere, on satellites or balloons. Such approach, however, limits the effective area of the detector, which in turn inhibits the measurements of low fluxes beyond the high energy (HE) range. In order to explore the more energetic gamma-rays, different detection approaches are needed—like the Imaging Air Cherenkov technique. Thanks to this technique, the ground-based telescopes can measure VHE photons indirectly, through the products of interactions induced by gamma-rays entering the atmosphere.

There are many different kinds of experiments (and approaches) used for direct and indirect detection of gamma-rays. However, as this work is about the VHE dark matter searches with Imaging Air Cherenkov Telescopes (IACTs), their detection technique is described in more detail. Space-based HE gamma-ray observatories are briefly covered as well, as their energy range (and results) complements those of the IACTs of the current generation.

For more complete review of other gamma-ray detection techniques and instruments, refer to e.g. [70].

2.3.4.1 Gamma-Ray Satellites

In the HE range, space-based observatories detect gamma-rays through the process of pair production: when an energetic photon penetrates the detector chamber, it interacts with matter creating an electron-positron pair. By measuring the paths and energies of these new particles, the properties of the original gamma-ray are reconstructed. Possible background from charged particles is usually reduced by anti-coincidence veto system surrounding the whole device.

The first gamma-ray satellite, Explorer 11 [71], was launched in 1961, but it was only a decade latter that the first detailed views of the gamma-ray sky were provided (Cos-B [72] and SAS-2 [73]). The first milestone in the field was set by the Energetic Gamma-Ray Experiment Telescope (EGRET, 1991-2000 [74]), that revealed more than 270 galactic and extragalactic objects radiating at energies between 100 MeV and 10 GeV. The most advanced pair-production gamma-ray detector of the current generation is the Large Area Telescope (LAT) on board the Fermi Gamma-ray Space Telescope (Fermi-GST) [75].

Fermi-GST satellite studies the gamma-ray sky with unprecedented sensitivity and angular resolution (Fig. 2.13). Launched in 2008, it resides in low-earth circular orbit, at altitude of 550 km, and operates in the all-sky survey mode (a whole sky sweep is performed every 3 h). Some of the scientific objectives the Fermi-GST was designed to fulfill include: understanding of acceleration mechanisms in active galactic nuclei, pulsars and supernova remnants; understanding of the gamma-ray diffuse emission; characterization of previously unidentified HE emitters, GRBs and other transient sources; search for dark matter signals and study of other questions of the Fundamental Physics.

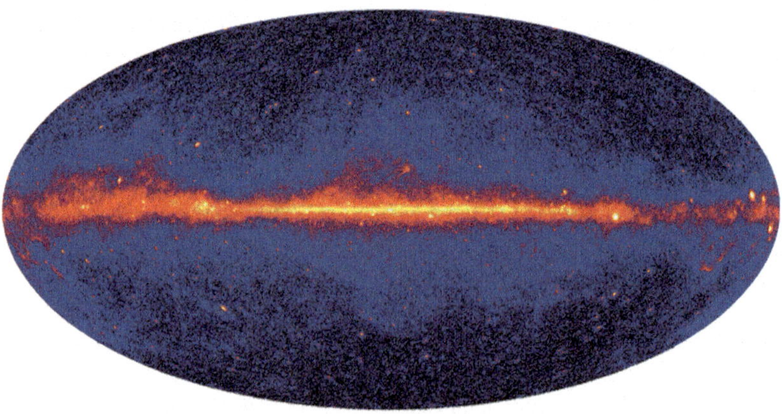

Fig. 2.13 Gamma-ray sky as seen by Fermi-LAT. The image is composed from 3 years of data, and it reveals bright emission in the plane of the Milky Way (*center*), bright pulsars and super-massive black holes. *Credit* NASA/DOE/International LAT Team

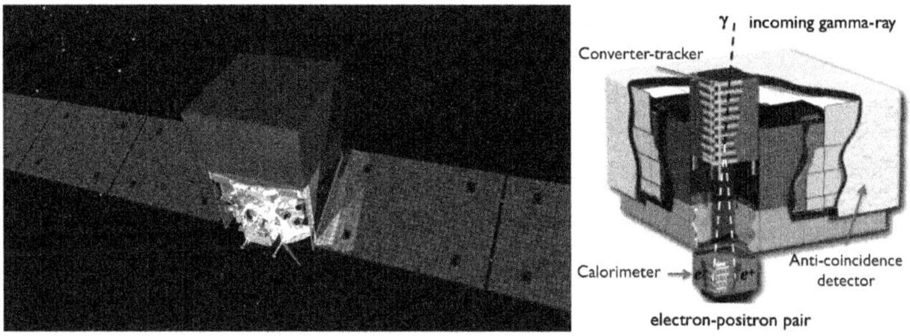

Fig. 2.14 *Left* simulation still of Fermi-GST in space. *Right* cutaway of the LAT instrument showing an inside view of one of the towers as an incoming gamma ray interacts producing an electron-positron pair. *Credit* NASA

The principal scientific component of the Fermi-GST is the already mentioned LAT [76], a pair-conversion instrument able to detect photons in the energy range from 30 MeV upto 300 GeV and with a field of view covering ∼20 % of the sky. The main background for Fermi-LAT are the charged cosmic rays, but an anti-coincidence detector, covering its top and lateral sides, allows for a 99.97 % background rejection (Fig. 2.14). As a result, Fermi-LAT has a very fruitful scientific production. Among other, a comprehensive list of gamma-ray sources (and their properties) detected by the LAT is published; the last edition, the Fermi-LAT Second Source catalog [77], contains 1873 sources, all of them characterized in the 100 MeV–100 GeV range. More than 500 of those are sources without a known counterpart at other wavelengths—the so-called UFOs (Sect. 2.3.3), some of which are potential dark matter targets. Fermi-LAT is contributing to the indirect dark matter searches in other ways as well: one of its first striking results was the discovery that active galactic nuclei are not the main source of the diffuse extragalactic background, as it was assumed, but instead can account for just a small fraction of it (∼16 %) [78]. This leaves room for an dark matter-based explanation for the dominant portion of the diffuse background, and there are numerous ongoing efforts in this direction. Thanks to its all-sky view, Fermi-LAT can accumulate great amounts of data from the most promising dark matter targets. For instance, this led to limits on dark matter parameter space from stacked galaxy clusters data [67] and from dSphs data [62]; more details of Fermi-LAT dark matter searches will be mentioned in Chap. 5.

For more details on Fermi satellite and its instruments, refer to [75].

2.3.4.2 Imaging Air Cherenkov Telescopes

When an energetic gamma-ray enters the atmosphere, it interacts with the nuclei of the medium creating an electron-positron pair that inherits its energy. The new particles then lose energy through the emission of bremsstrahlung photons, that in turn create more electron-positron pairs and so on. The process repeats itself, with pairs less energetic in every step, until the probability for bremsstrahlung emission becomes smaller than the energy loss through ionization (which becomes the case for critical energy below 83 MeV). This cascade of particles, originated by the initial gamma-ray, is called the extensive atmospheric shower (EAS). It develops over hundreds of meters in width and several kilometers in length, with a maximum of produced particles typically at altitude of 8–12 km a.s.l. In general, EAS can be induced by different kinds of particles; as in the ones initiated by gamma-rays only photons, electrons and positrons take part, such EAS is called electromagnetic (EM) shower. A large fraction of charged particles of the cascade move faster than light in the atmosphere, producing Cherenkov radiation—a 'bluish' light emitted within a conic structure, starting at the EAS and ending on the ground in a circle of $\sim$120 m in radius (Fig. 2.15). If a reflector is located within this Cherenkov light pool, it can collect a part of this light and focus it onto the extremely sensitive and fast camera, composed of high efficiency photodetectors, capable of resolving the image of the shower. From there, by employing the image reconstruction algorithms, the arrival direction and energy of the primary gamma-ray are determined. If two or more of these IACTs record the same event, a stereo image of the shower is obtained, allowing for a better reconstruction of the primary photon. The main issue for this approach is the dominant background, mainly produced by *hadronic showers* (Fig. 2.16). These EAS, initiated by protons and heavier nuclei, outnumber the showers created by gamma-rays by order of $\sim$10000. Population of these EAS includes neutral pions that immediately decay into a pair of gamma-rays, thus consequently generating secondary EM showers, and the Cherenkov photons that, when recorded, show the image of the hadronic EAS in the camera. The morphology of such image differs from the one of the EM shower (Fig. 2.17), and this property is used to discriminate between these two populations of events. Selection algorithms can reject more than 99 % of hadron-induced showers.

Other kinds of unwanted background events that can trigger an IACT include:

- *electron-induced EM showers*: electrons from the cosmic-ray population can generate purely EM cascades, thus undistinguishable from the gamma-ray induced ones. Consequently, this background cannot be suppressed. On the up side, electron flux (above 100 GeV) is approximately isotropic and much smaller than the gamma-ray flux, which strongly reduces the impact of this background on the observations;
- *diffuse gamma-ray emission*: irreducible and unavoidable background. Strictly speaking, it is not inherent to the method, but to the physics case itself; it is also present in the HE regime. It consists of two components: the extragalactic and the galactic gamma-ray background. Fortunately, neither one is too significant at VHE;
- *muon-induced Cherenkov photons*: muons are produced in hadronic showers, and usually reach the ground before decaying into electron and the corresponding neutrino. Muons do not generate EM showers, but they can radiate Cherenkov photons. If this occurs at moderate altitudes, it mimics the image of a low energy EM shower, while muons reaching the ground near the telescope produce the

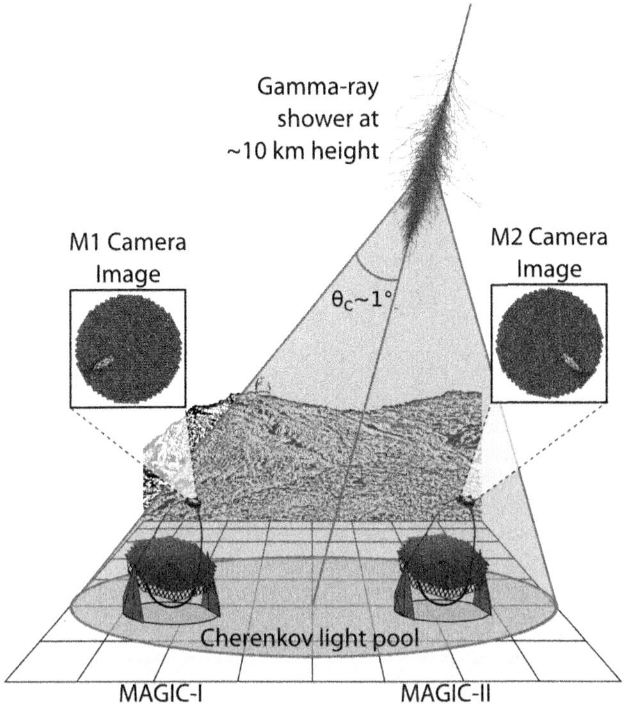

Fig. 2.15 Sketch of the principle of the Imaging Air Cherenkov technique, through the formation of the image of an EAS in an IACT pixelated camera. The numbers in the Figure correspond to a typical 1 TeV gamma-ray induced shower. *Credit* G. Giavitto

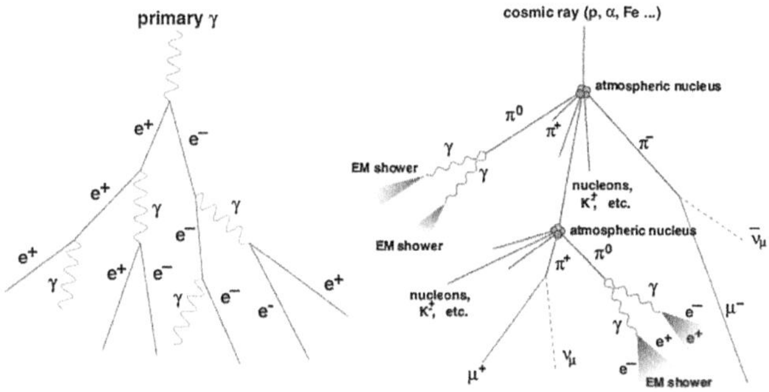

Fig. 2.16 Sketch of the structure and the interactions present in an EAS, induced by a γ-ray (*left*) and by a hadron (*right*)

so-called muon rings in the camera. The incidence of either case is highly suppressed in the stereoscopic observations by the offline data analysis;

- *Night-sky background* (NSB) *light*: NSB is ambient light, produced by the diffuse scattered light from the stars, the Moon, or human activities in the vicinity of

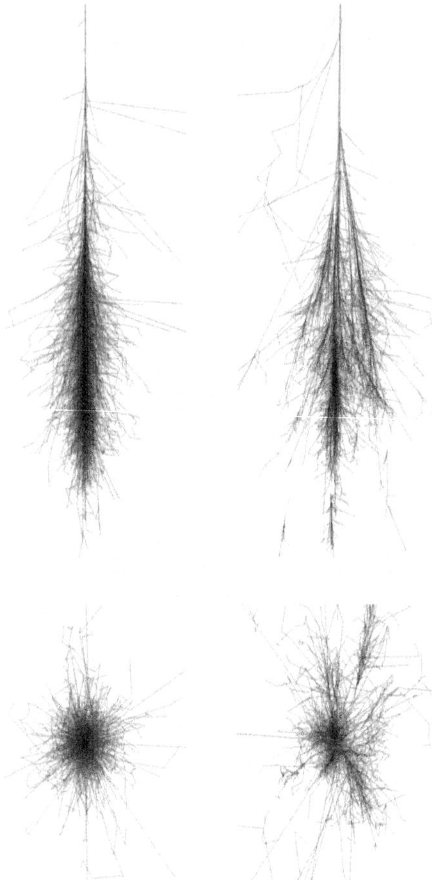

Fig. 2.17 Longitudinal (*top*) and lateral (*bottom*) development of an electromagnetic (*left*) and hadronic (*right*) showers with an initial energy of E = 100 GeV simulated with CORSIKA. The longitudinal projection plot has a height of 20 km and a width of 10 km, while the lateral projection plot is 10 km on the side. Taken from [79]

the telescope. Fluctuations in NSB are reflected as small images in the camera, in a way that mimics the low energy gamma-ray events. NSB is amplified in the presence of clouds or air pollution;

- *electronic noise*: the camera pixels and the readout chain of the telescope produce an intrinsic electronic noise that must be taken into account in the data analysis. Under certain circumstances, high electronic noise can even trigger the instrument as if it was a low-energy gamma-ray shower. To suppress this kind of background, it is needed to have it well characterized.

The Imaging Air Cherenkov technique is relatively young, compared to other astronomical methods employed in radio and optical telescopes. The first ever detection of a VHE gamma-ray source occurred in 1989—the Crab Nebula was observed by the pioneering, 10 m diameter Whipple telescope [80]. In the years that followed, the technique was successfully used by other experiments, like HEGRA [81] and CAT [82], increasing the number of detected VHE emitters to $\sim$10 (mainly nearby active galactic nuclei). The current generation of IACTs, represented by MAGIC, H.E.S.S. and VERITAS (Very Energetic Radiation Imaging Telescope Array System, [83]), is marked by significant technological improvements with respect to the previous one—this is reflected in an order of magnitude better flux sensitivity and greatly reduced energy threshold. Consequently, the number of detected VHE sources, from both galactic and extragalactic populations, has surpassed the number of 100. The next step in the field is the construction of the Cherenkov Telescope Array (CTA), observatory of 10 times better sensitivity, 10-fold wider energy range and increased capability for physics achievements with respect to the current IACTs (Sect. 6.1, [84]).

The results presented in this work are based on the observations performed with the MAGIC Telescopes. The following chapter gives details on the technical properties and analysis procedures of MAGIC.

References

1. M. Battaglia, M.E. Peskin, Supersymmetric dark matter at colliders, in *Particle Dark Matter: Observations, Models and Searches*, ed. by G. Bertone, 1st edn. (Cambridge University Press, Cambridge, 2010), pp. 276–305. ISBN: 978-0521763684
2. CERN European Organization for Nuclear Research: The ATLAS experiment home page (2013), http://atlas.web.cern.ch/Atlas/Collaboration
3. CERN European Organization for Nuclear Research: The CMS experiment home page (2013), http://cms.web.cern.ch
4. K. Kowalska, L. Roszkowski, E.M. Sessolo, Two ultimate tests of constrained supersymmetry (2013), arXiv:1302.5956
5. ATLAS Collaboration: Search for strongly produced supersymmetric particles in final states with two same-sign leptons and jets with the ATLAS detector using 21 fb^{-1} of proton-proton collisions at $\sqrt{s} = 8$ TeV, ATLAS note ATLAS-CONF-2013-007, CERN, Geneve, Switzerland (2013)
6. CERN Courier: Work for the LHCs first long shutdown gets under way (2013)
7. S. Chatrchyan et al., Search for dark matter and large extra dimensions in monojet events in pp collisions at $\sqrt{s} = 7$ TeV. JHEP **09**, 094 (2012), arXiv:1206.5663
8. D.G. Cerdeño, A.M. Green, Direct detection of WIMPs, in *Particle Dark Matter: Observations, Models and Searches*, ed. by G. Bertone, 1st edn. (Cambridge University Press, Cambridge, 2010), pp. 347–369. ISBN: 978-0521763684
9. R. Bernabei et al., First results from DAMA/LIBRA and the combined results with DAMA NaI. Eur. Phys. J. C **56**, 333–355 (2008), arXiv:0804.2741
10. C.E. Aalseth et al., Results from a search for light-mass dark matter with a P-type point contact germanium detector. Phys. Rev. Lett. **106**, 31301 (2011), arXiv:1002.4703
11. The XENON DarkMatter Project: XENON100 experiment (2013), http://xenon.astro.columbia.edu/XENON100_Experiment
12. E. Aprile et al., Dark matter results from 225 live days of XENON100 data. Phys. Rev. Lett. **109**, 181301 (2012), arXiv:1207.5988
13. E. Aprile et al., Limits on spin-dependent WIMP-nucleon cross sections from 225 live days of XENON100 data (2013), arXiv:1301.6620
14. E. Armengau et al., Radiopure Scintillators for EURECA, in *Proceedings of the 1st International Workshop RPScint*, Kyiv, Ukraine (2009), arXiv:0903.1539
15. L. Baudis, DARWIN: dark matter WIMP search with noble liquids, in *Proceedings of the Identification of Dark Matter*, Montpellier, France (2010), arXiv:1012.4764

16. E. Aprile et al., Performance of a cryogenic system prototype for the XENON1T Detector. J. Instrum. **7**, 10001 (2012), arXiv:1208.2001
17. The ANTARES Collaboration: ANTARES home page (2013), http://antares.in2p3.fr/
18. University of Wisconsin-Madison: IceCube neutrino observatory (2013), http://icecube.wisc.edu
19. F. Halzen, D. Hooper, High-energy neutrinos from WIMP annihilations in the Sun, in *Particle Dark Matter: Observations, Models and Searches*, ed. by G. Bertone, 1st edn. (Cambridge University Press, Cambridge, 2010), pp. 507–520. ISBN: 978-0521763684
20. R. Abbasi et al., Search for neutrinos from annihilating dark matter in the direction of the Galactic Center with the 40-string IceCube neutrino observatory (2012), arXiv:1210.3557
21. PAMELA Mission: A Payload for Antimatter Exploration and Light-nuclei Astrophysics Mission official website (2013), http://pamela.roma2.infn.it
22. O. Adriani et al., An anomalous positron abundance in cosmic rays with energies 1.5–100 GeV. Nature **458**, 607–609 (2009), arXiv:0810.4995
23. I. Moskalenko, A. Strong, Production and propagation of cosmic-ray positrons and electrons. Astrophys. J. **493**, 694–707 (1998), arXiv:astro-ph/9710124
24. A.A. Abdo et al., Measurement of the cosmic ray e^+ plus e^- spectrum from 20 GeV to 1 TeV with the Fermi Large Area Telescope. Phys. Rev. Lett. **102**, 181101 (2009), arXiv:0905.0025
25. The AMS-02 Collaboration: The Alpha Magnetic Spectrometer experiment (2013), http://www.ams02.org
26. M. Aguilar et al., First result from the Alpha Magnetic Spectrometer on the International Space Station: precision measurement of the positron fraction in primary cosmic rays of 0.5350 GeV. Phys. Rev. Lett. **110**, 141102 (2013)
27. M. Cirelli et al., Model-independent implications of the e^+, e^-, anti-proton cosmic ray spectra on properties of dark matter. Nucl. Phys. B **813**, 1–21 (2009), arXiv:0809.2409
28. A. Ibarra, D. Tran, C. Weniger, Decaying dark matter in light of the PAMELA and Fermi LAT data. JCAP **01**, 009 (2010), arXiv:0906.1571
29. D. Hooper, P. Blasi, P. Dario Serpico, Pulsars as the sources of high energy cosmic ray positrons, JCAP **01**, 025 (2009), arXiv:0810.1527
30. O. Adriani et al., A newmeasurement of the antiproton-to-proton flux ratio up to 100 GeV in the cosmic radiation. Phys. Rev. Lett. **102**, 051101 (2009), arXiv:0810.4994
31. P. Salati, F. Donato, N. Fornengo. Indirect dark matter detection with cosmic antimatter, in *Particle Dark Matter: Observations, Models and Searches*, ed. by G. Bertone, 1st edn. (Cambridge University Press, Cambridge, 2010), pp. 521-546. ISBN: 978-0521763684
32. A. Ibarra, D. Tran, C. Weniger, Detecting gamma-ray anisotropies from decaying dark matter: prospects for Fermi LAT. Phys. Rev. D **81**, 023529 (2010), arXiv:0909.3514
33. J.F. Navarro, C.S. Frenk, S.D. White, The structure of cold dark matter halos. Astrophys. J. **462**, 563–575 (1996), arXiv:astro-ph/9508025
34. C.A. Vera-Ciro et al., Not too big, not too small: the dark halos of the dwarf spheroidals in the Milky Way. Mon. Not. Roy. Astron. Soc. **428**, 1696–1703 (2013), arXiv:1202.6061
35. J.F. Navarro et al., The diversity and similarity of simulated cold dark matter haloes. Mon. Not. Roy. Astron. Soc. **402**, 21–34 (2010), arXiv:0810.1522
36. A.V. Kravtsov et al., The cores of dark matter dominated galaxies: theory vs. observations. Astrophys. J. **502**, 48–58 (1998), arXiv:astro-ph/9708176
37. N.W. Evans, J. An, M.G. Walker, Cores and cusps in the dwarf spheroidals. Mon. Not. R. Astron. Soc. Lett. **393**, L50–L54 (2009), arXiv:0811.1488
38. V. Springel et al., The Aquarius project: the subhaloes of galactic haloes. Mon. Not. R. Astron. Soc. **391**, 1685–1711 (2008), arXiv:0809.0898
39. J. Diemand, M. Kuhlen, P. Madau, Dark matter substructure and gamma-ray annihilation in the Milky Way Halo. Astrophys. J. **657**, 262–270 (2007), arXiv:astro-ph/0611370
40. L.E. Strigari et al., Precise constraints on the dark matter content of Milky Way dwarf galaxies for gamma-ray experiments. Phys. Rev. D **75**, 083526 (2007), arXiv:astro-ph/0611925
41. L.E. Strigari, Galactic searches for dark matter (2012), arXiv:1211.7090
42. E.W. Kolb, M.S. Turner, The early Universe. Front. Phys. **69**, 1 (1990)

43. J. Hisano, S. Matsumoto, M.M. Nojiri, Explosive dark matter annihilation. Phys. Rev. Lett. **92**, 031303 (2004), arXiv:hep-ph/0307216
44. P.A.R. Ade et al., Planck 2013 results. XVI, Cosmological parameters (2013), arXiv:1303.5076
45. G. Bertone et al., Gamma-rays from decaying dark matter. JCAP **11**, 003 (2012), arXiv:0709.2299
46. J. Cembranos et al., Photon spectra from WIMP annihilation. Phys. Rev. D **83**, 083507 (2011), arXiv:1009.4936
47. T. Bringmann et al., Fermi LAT search for internal bremsstrahlung signatures from dark matter annihilation. JCAP **07**, 054 (2012), arXiv:1203.1312
48. A. Birkedal et al., Robust gamma-ray signature of WIMP dark matter (2005), arXiv:hep-ph/0507194
49. T. Bringmann, L. Bergström, J. Edsjö, New gamma-ray contributions to supersymmetric dark matter annihilation. JHEP **01**, 049 (2008), arXiv:0710.3169
50. A. Ibarra, S. Lopez Gehler, M. Pato, Dark matter constraints from box-shaped gamma-ray features. JCAP **07**, 043 (2012), arXiv:1205.0007
51. D. Merritt et al., Dark matter spikes and annihilation radiation from the Galactic Center. Phys. Rev. Lett. **88**, 191301 (2002), arXiv:astro-ph/0201376
52. F. Aharonian et al., H.E.S.S. observations of the Galactic Center region and their possible dark matter interpretation. Phys. Rev. Lett. **97**, 221102 (2006), arXiv:astro-ph/0610509
53. J. Albert et al., Observations of gamma rays from the Galactic Center with the MAGIC telescope. Astrophys. J. Lett. **638**, L101–L104 (2006), arXiv:astro-ph/0512469
54. The H.E.S.S. Collaboration: H.E.S.S.—the High Energy Spectroscopic System (2013), http://www.mpi-hd.mpg.de/hfm/HESS
55. A. Abramowski et al., Search for a dark matter annihilation signal from the Galactic Center halo with H.E.S.S. Phys. Rev. Lett. **106**, 161301 (2011), arXiv:1103.3266
56. C. Weniger, A tentative gamma-ray line from dark matter annihilation at the Fermi Large Area Telescope. JCAP **08**, 007 (2012), arXiv:1204.2797
57. A. Ibarra et al., Gamma-ray boxes from axion-mediated dark matter. JCAP **05**, 016 (2013), arXiv:1303.6632
58. W. Buchmuller, M. Garny, Decaying vs annihilating dark matter in light of a tentative gamma-ray line. JCAP **08**, 035 (2012), arXiv:1206.7056
59. A. Hektor, M. Raidal, E. Tempel, Double gamma-ray lines from unassociated Fermi-LAT sources revisited (2012), arXiv:1208.1996
60. M. Ackermann et al., Search for gamma-ray spectral lines with the Fermi Large Area Telescope and dark matter implications (2013), arXiv:1305.5597
61. D. Nieto, Dark matter constrains from high energy astrophysical observations. Ph.D. thesis. Universidad Complutense de Madrid, Spain (2012)
62. M. Ackermann et al., Constraining dark matter models from a combined analysis of Milky Way satellites with the Fermi Large Area Telescope. Phys. Rev. Lett. **107**, 241302 (2011), arXiv:1108.3546
63. L.E. Strigari et al., The most dark-matter-dominated galaxies: predicted gamma- ray signals from the faintest Milky Way dwarfs. Astrophys. J. **678**, 614–620 (2008),arXiv:0709.1510
64. N. Evans, F. Ferrer, S. Sarkar, A baedecker for the dark matter annihilation signal. Phys. Rev. D **69**, 123501 (2004), arXiv:astro-ph/0311145
65. A. Pinzke, C. Pfrommer, L. Bergström, Prospects of detecting gamma-ray emission from galaxy clusters: cosmic rays and dark matter annihilations. Phys. Rev. D **84**, 123509 (2011), arXiv:1105.3240
66. M.A. Sanchez-Conde et al., Dark matter searches with Cherenkov telescopes: nearby dwarf galaxies or local galaxy clusters? JCAP **12**, 011 (2011), arXiv:1104.3530
67. M. Ackermann et al., Constraints on dark matter annihilation in clusters of galaxies with the Fermi Large Area Telescope. JCAP **05**, 025 (2010), arXiv:1002.2239
68. J. Aleksić et al., MAGIC gamma-ray telescope observation of the Perseus cluster of galaxies: implications for cosmic rays, dark matter, and NGC 1275. Astrophys. J. **710**, 634–647 (2010), arXiv:0909.3267

69. A. Boyarsky et al., Restrictions on parameters of sterile neutrino dark matter from observations of galaxy clusters. Phys. Rev. D **74**, 103506 (2006), arXiv:astro-ph/0603368
70. V. Schönfelder (ed.) *The Universe in Gamma Rays* (Springer, Hamburg, 2001). ISBN: 978-3540678748
71. NASA Goddard Space Flight Center: The Explorer-11 Satellite (2013), http://heasarc.gsfc.nasa.gov/docs/heasarc/missions/explorer11.html
72. European Space Agency: Cos-B overview (2013), www.esa.int/Our_Activities/Space_Science/Cos-B_overview2
73. NASA Goddard Space Flight Center: The Small Astronomy Satellite 2 (2013), http://heasarc.gsfc.nasa.gov/docs/sas2/sas2.html
74. NASA Goddard Space Flight Center: The Energetic Gamma Ray Experiment Telescope (EGRET) (2013), http://heasarc.gsfc.nasa.gov/docs/cgro/egret
75. NASA Goddard Space Flight Center: The Fermi Gamma-Ray Space Telescope (2013), http://fermi.gsfc.nasa.gov
76. W.B. Atwood et al., The Large Area Telescope on the Fermi Gamma-Ray Space Telescope mission. Astrophys. J. **697**, 1071–1102 (2009), arXiv:0902.1089
77. P.L. Nolan et al., Fermi Large Area Telescope second source catalog. Astrophys. J. Supp. **199**, 46 (2012), arXiv:1108.1435
78. A.A. Abdo et al., The Fermi-LAT high-latitude survey: source count distributions and the origin of the extra galactic diffuse background. Astrophys. J. **720**, 435–453 (2010), arXiv:1003.0895
79. F. Schmidt, CORSIKA shower images (2013), http://www.ast.leeds.ac.uk/~fs
80. T.S. Weeks et al., Observations of TeV gamma rays from the Crab nebula using the atmospheric Cherenkov imaging technique. Astrophys. J. **342**, 379–395 (1989)
81. Max-Planck-Institut für Kernphysik: The HEGRA atmospheric Cherenkov telescope system (2013), http://www.mpi-hd.mpg.de/hfm/CT
82. A. Barrau et al., The CAT imaging telescope for very-high-energy gamma-ray astronomy. Nucl. Instrum. Meth. A **416**, 278–292 (1998), arXiv:astro-ph/9804046
83. The VERITAS Collaboration: VERITAS home page (2013), http://veritas.sao.arizona.edu
84. The CTA Consortium: The Cherenkov Telescope Array home page (2013), https://www.cta-observatory.org

Chapter 3
The MAGIC Telescopes

The *Florian Goebel* Major Atmospheric Gamma-ray Imaging Cherenkov (MAGIC) Telescopes [1] are located at the Roque de los Muchachos Observatory (28.8° N, 17.9° W; 2200 m a.s.l.) in the Canary island of La Palma. The system consists of two, 17 m diameter telescopes (Fig. 3.1), designed to have high sensitivity at low energies and fast response to variable phenomena. The first instrument, MAGIC-I, has been operational since 2004. In 2009, the second telescope, MAGIC-II, was constructed and commissioned, and the system has been successfully running in the stereoscopic mode ever since. During 2011 and 2012, the telescopes underwent a major upgrade, aimed at homogenization and improvement of the performance of both instruments.

This chapter is devoted to the description of technical properties of the MAGIC Telescopes, as well as of the standard observational procedures and of the analysis chain used to process the gathered data.

3.1 Technical Description

The principles on which MAGIC Telescopes were constructed include large and light-weighted reflector, quick repositioning, prompt and accurate tracking of any source on the sky, fast electronics and highly sensitive sensors able to catch even the faintest flashes of Cherenkov light. As a result, the system can explore the lowest regions of the VHE band, between 30 and 100 GeV, with energy threshold of 50 GeV for the standard observations and the integral sensitivity of $\sim$0.7 % of the Crab Nebula flux, above few hundred GeV, for a 5σ detection in 50 h [2]. Moreover, it can move to any position in only tens of seconds, a property that is of great importance when trying to detect extremely short gamma-ray bursts (GRBs). This section gives some details on technical characteristic of particular subsystems of MAGIC.

J. Aleksić, *Optimized Dark Matter Searches in Deep Observations of Segue 1 with MAGIC*, Springer Theses, DOI 10.1007/978-3-319-23123-5_3

Fig. 3.1 The MAGIC Telescopes, Roque de los Muchachos Observatory, La Palma. *Credit* the MAGIC Collaboration

3.1.1 Mount, Drive and Mirrors

Mount The mounting structures of MAGIC telescopes, identical for both instruments (Fig. 3.2), are made so that they are large, light-weighted and rigid. Large frame is needed for hosting the 17 m reflector, which in turn permits low energy threshold. The frame consists of a three-layer structure made of light, carbon fiber tubes, that weight about 5.5 tons—one third of the conventional steel-made structure. This permits fast movement. The structure material is also quite resistant to deformations, and as such it prevents significant reflector alterations and worsening of the image quality. Still, some structure bending is inevitable: the camera, located at a focal distance of $\sim$17 m, carried by a single tubular arch and stabilized by thin steel cables anchored to the main dish frame, weights more than half a ton, therefore straining the structure during the telescope tracking. However, the mount deformation is less than 3.5 mm for any orientation of the telescope [3], and its effect on the image can be corrected by an automatic system of mirror re-orientation—Active Mirror Control (AMC).

Drive The mount of the telescope uses Alt-Azimuth drive to track objects. The continuous observation of a source, without reaching any end position in either altitude or azimuth coordinates, is possible given the wide range of allowed movements: from $-90°$ to $+318°$ in azimuth, and from $-70°$ to $+105°$ in elevation. Motion is powered by two servo-motors for azimuthal and one servo-motor for the elevation axis (Fig. 3.2), and it is limited by the mechanical end-switches [3]. In speed GRB mode, the telescope can turn 180° in azimuth in $\sim$20 s. The pointing of the system is constantly cross-checked by three absolute 14-bit shaft encoders: during normal observations, the intrinsic mechanical accuracy of the pointing system is much better

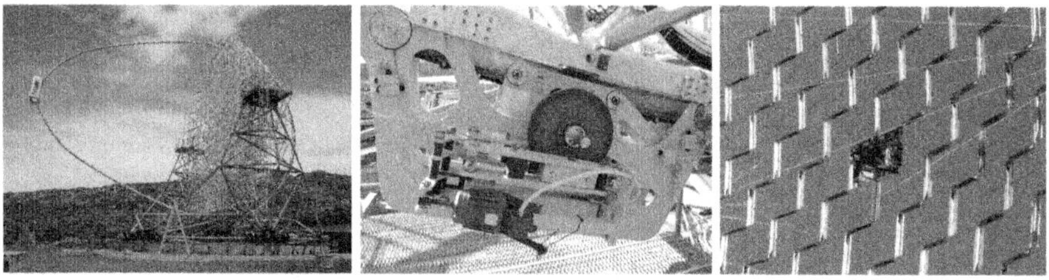

Fig. 3.2 Frame structure of MAGIC-I with arcs holding the camera (*left*). Elevation motor (*center*). MAGIC-I mirror staggering (*right*). *Credit* The MAGIC Collaboration

than $1'$ on the sky (1/5 of a pixel diameter). In addition, the pointing precision of the telescope is monitored by a system called Starguider, that allows further, offline correction of mispointing via software. The *Starguider* system consists of a $4.6°$ field of view (FoV) CCD camera installed close to the center of the reflector, that observes the camera of the telescope and determines the exact position of its center by the means of LEDs, placed on the telescope camera frame, and the part of the sky close to the telescope camera (chosen so it contains bright, reference star). Possible offsets are calculated through the comparison of the observed and catalog coordinates of the "guide" stars, thus inferring the actual pointing of the telescope. The reliability of such mispointing measurement depends on the sky visibility (atmospheric conditions), and is estimated using the ratio of the number of observed and expected stars.

Mirrors The 17 m diameter reflector follows a parabolic profile which allows preservation of the temporal structure of the Cherenkov photons. The benefit of this shape is twofold: it reduces the time window required for signal extraction, therefore reducing the integrated noise, and it permits the use of time evolution as a discrimination parameter between hadronic and electromagnetic showers (seen how development of the hadronic showers takes longer, Sect. 2.3.4.2).

MAGIC-I reflector consists of 956 $0.5 \times 0.5\ m^2$ aluminum honeycomb mirrors of high reflectivity (80–90 %), for a total surface of $239\ m^2$. The reflector of MAGIC-II is built from 143 full-aluminium and 104 glass-aluminium mirrors, with each panel being of $1\ m^2$ in size, for a total surface of $247\ m^2$ [4]. Each mirror can be oriented by the AMC depending on the telescope elevation.

3.1.2 Camera and Calibration System

When light hits the mirrors of the telescope, part of it is reflected and focused into the camera. The camera is one of the crucial systems of the telescope, as its performance conditions the sensitivity, energy threshold and signal/background discrimination capacity of the detector. The camera hosts clusters of photo-multiplier tubes (PMTs), that convert photons into electric signals which are then carried to the telescope

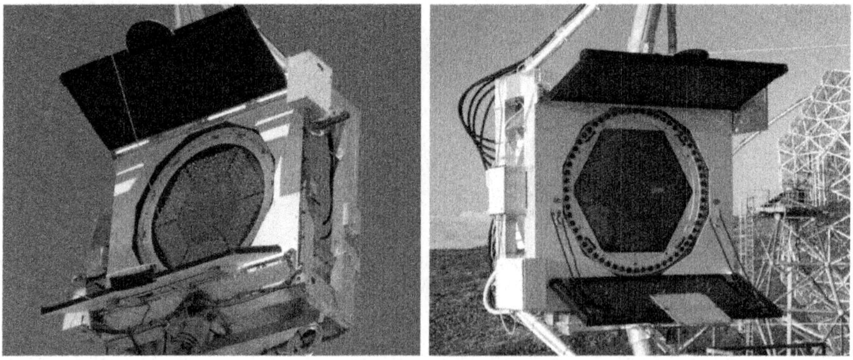

Fig. 3.3 *Left* Front of the MAGIC-I camera (before the upgrade) with two types of PMTs visible. *Right* Closeup of the MAGIC-II camera. *Credit* The MAGIC Collaboration

readout. The key features for PMTs selection were high quantum efficiency (QE), fast response, low gain and fine pixelization. The photon entrance of each PMT is equipped with a hexagonal light collector (Winston cone), that increases the entry window for each pixel as well as the double-crossing probability of photons with large incidence angles. Furthermore, Winston cone prevents stray light (not coming from the reflector) from entering the PMT. The PMTs are double-protected by a transparent plexiglas window and by two lids that open only during the observations (Fig. 3.3). In addition, the camera is equipped with an air cooling and heating systems, preventing overheating or moisture accumulation.

Although MAGIC-II was constructed as a mechanical clone of MAGIC-I, the camera of the former was built with different structure and using components of improved performance. Its efficiency was cause for an upgrade of the MAGIC-I camera with an exact copy of the MAGIC-II camera, action successfully completed in Summer 2012. As the observations analyzed in this work have been recorded with the both old and new MAGIC-I camera, details are given for both the pre- and post-upgrade systems.

MAGIC-I Camera (Pre-Upgrade) The old MAGIC-I camera was of hexagonal shape, and $\sim$3.6° FoV (Fig. 3.3). It was equipped with 577 PMTs of two types: the inner section of the camera (up to 2.1° diameter) was made of 397 pixels of 30 mm diameter and 0.1° FoV. This region (approximately) represented the trigger area. The outer region was composed of the remaining 180 PMTs, of 60 mm diameter and 0.2° FoV. The hemispherical PMT photocathode was coated with a wavelength shifter that was decreasing the Cherenkov photon frequency and increasing the peak QE (up to 28 %). The central pixel of the MAGIC-I camera was specially designed to perform optical measurements (mainly the optical pulsations of the Crab Nebula pulsar) to check the timestamp of the overall system.

The calibration of the old MAGIC-I camera was performed by an optical calibration system installed in the center of the reflector. The system employed ultra-fast LEDs (that emulate the fast timing behavior of the Cherenkov pulses) with different characteristic wavelengths (370, 460 and 520 nm), able to uniformly illuminate the camera with light pulses of custom frequency and intensity.

MAGIC-II and the Post-Upgrade MAGIC-I Camera This camera is of roughly circular shape, with 1.2 m diameter and 3.5° FoV (Fig. 3.3). It is composed of 1039 PMTs of 30 mm diameter and 0.1° FoV each, grouped in 169 independent clusters. The trigger area covers the innermost 2.5° diameter region.

The calibration system is installed in the center of the telescope reflector. It consists of a frequency tripled passively Q-switched Nd-YAG laser, whose beam passes through two filter wheels with different attenuation factors, allowing for easy adjustment of the pulse intensity. After the attenuation, the laser beam is diffused via an integrating (Ulbricht) sphere, providing a homogeneous illumination of the camera.

3.1.3 Readout System

Following the conversion of Cherenkov photons into electric pulses in the PMTs, the electric signal is amplified and converted into optical signal by the vertical cavity surface laser diodes (VCSELs), located at the base of each PMT. The VCSELs are coupled to 162 m of optical fibers that transmit the signal to the *Counting House*, a building hosting the rest of the electronics and the operations center. This way of analog optical transmission protects the signal from ambient electromagnetic noise in the line, preserves the bandwidth and reduces the signal losses from the camera to the readout. In the Counting House, optical signals are converted back into electric pulses in the receiver boards and split into two branches, by means of fast GaAs PIN diodes. One branch goes into the trigger system, while the other is passed to the data acquisition system (DAQ). The DAQ, through the use of Analog to Digital Converters (ADCs), digitizes the electric pulses at a speed of 2 GSamples/s. When trigger arrives, the sampling stops and the digitized signal is recorded into raw data file.

When MAGIC-II was built, its readout system differed from that of MAGIC-I. In Autumn 2011, a partial upgrade of the telescopes was performed, including replacement of both readout systems with the new ones, identical for the two instruments. As this change affects the observations presented in this work, the pre-upgrade as well as the post-upgrade readout systems are described in detail below.

3.1.3.1 Pre-Upgrade

MAGIC-I When the optical signal reaches the MAGIC-I readout, it is split in two branches before reaching the receiver boards. One half of the signal is sent to the trigger, where it is converted back into an analog electric pulse. The other half is forwarded to the fiber-optic multiplexing readout system (MUX) [5], where pulses from every 16 channels are linked together. This is done by delaying the optical signal of each channel by multiples of 40 ns before converting them back into electric pulses by means of fast GaAs PIN diodes. The signal is then digitized by 2 GSamples/s Flash ADCs and written to a ring buffer. When trigger is issued, the corresponding part of the buffer is written to the disk. The bandwidth of the whole DAQ chain is about 250 Hz, and the dead time is $\sim$25 μs.

MAGIC-II The readout of MAGIC-II first converts the optical back into electric signal in the high bandwidth and fully programmable receiver boards called MONSTER (Magic Optical NanoSecond Trigger and Event Receiver), and then splits the signal into two branches. One branch is sent to a discriminator with a software adjustable threshold, and the signal that surpasses the chosen threshold is digitized and routed to the trigger system. The other branch of the signal is passed to the digitizing units. The core of the MAGIC-II DAQ is an ultra fast analog sampler, called Domino Ring Sampler, version 2 (DRS2) [6]. A single Domino chip hosts 10 input analog channels, each associated to one pixel in the camera. One DRS2 channel is equipped with 1024 capacitive cells, organized as a ring buffer where signal is continuously recorded. When trigger arrives, the sampling is stopped and the signal currently stored in the capacitors is frozen, then read out and digitized at 40 MHz rate. Each time, all of the 1024 capacitors have to be read out, bringing the dead time of the system to 0.5 ms. DRS2 chips are installed in pairs on custom made mezzanine cards, that in turn are mounted (in groups of 4) on the PULSAR (PULSer And Recorder) boards. PULSAR reads out the analog input signal which is then transmitted to the HOLA (High Optical Link for Atlas) board, and from there forwarded via optical S-Link to the FILAR (Four Input Links for Atlas Readout) board. FILARs are installed in DAQ pc that interfaces with user and where the readout system ends. The schematic representation of a fully equipped analog PULSAR is shown on Fig. 3.4. In addition to 14 analog PULSAR boards, there are two special PULSARs that belong to the DAQ system: the DIGITAL PULSAR, implemented to add digital information to the data (such as absolute time and the trigger number), and BUSY PULSAR that stops the triggers when the readout is already processing one event or when an error occurs.

3.1.3.2 Post-Upgrade

By the end of 2011, the readouts of MAGIC-I and MAGIC-II have been replaced by upgraded systems, largely based on the previous DRS2 readout [7]. The main change is the use of an more sophisticated digitizing unit, DRS4. The new chip has several important advantages over DRS2: a completely linear behavior as well as the option to operate in the limited, region-of-interest (ROI) readout mode, which significantly reduces the dead time (to $\sim$26 μs).

3.1.4 Trigger

The trigger system discriminates Cherenkov showers from the NSB, by selecting only fast pulses (<5 ns) detected simultaneously in several neighboring pixels. Trigger is composed of four different trigger levels: the first three work over the individual telescope data, while the fourth is shared by both instruments.

- Level 0 (L0): hosted on the receiver boards, it evaluates every channel individually and issues a trigger only if it exceeds a certain amplitude—the discriminator threshold (DT). DTs are programmable by software.
- Level 1 (L1): L0 signals are sent to L1, which examines the channels in search for spatial and temporal correlation over the decomposition of the trigger region in 19 overlapping macrocells (Fig. 3.5). This topological condition is based on the close compact next neighbor (CCNN) logic, that is, only events present in N adjacent pixels are accepted by the L1 trigger. The standard CCNN setup applied in MAGIC observations is that of 4NN, with 2NN, 3NN and 5NN also available.
- Level 2 (L2): in the first version of the MAGIC-I trigger, events triggered by L1 entered L2 for further, shape-based discrimination. L2 was never enabled for standard data taking, but its unit was used for event rate monitoring, rate scaling and integration of L1, calibration and stereo triggers.
- Level 3 (L3—Stereo trigger): selects only those events that have triggered both telescopes individually within a certain time interval. In order to minimize the coincidence time window, the arrival times of L1 triggers of each telescope are delayed, depending on the pointing positions of the telescopes.

Aside from these trigger levels that are considered standard in MAGIC observations, there is another trigger concept—the *sum trigger*—specifically designed to lower the trigger threshold by a factor of 2, down to ~25 GeV. The sum trigger logic is based on sum of several analog signals, in such a way that the signal-to-noise ratio of low energy showers is minimized over the NSB light. The sum trigger was

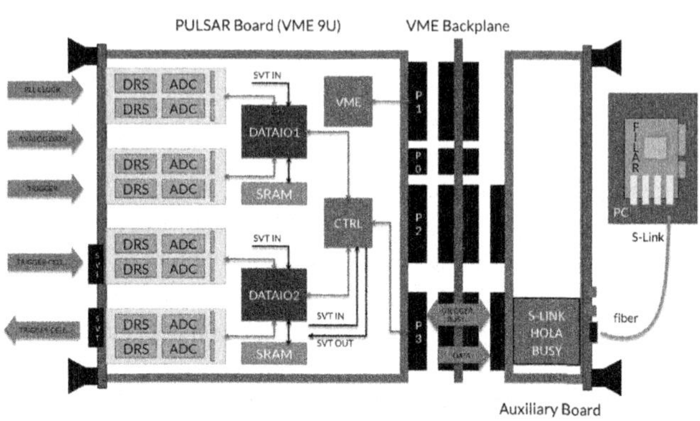

Fig. 3.4 Partial schematics of the MAGIC-II pre-upgrade DAQ system

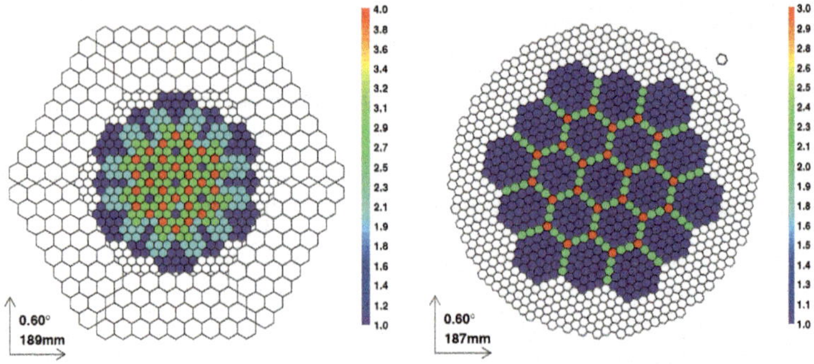

Fig. 3.5 Schematic representation of the MAGIC-I (pre-upgrade, *left*) and MAGIC-II and post-upgrade MAGIC-I (*right*) cameras with L1 trigger areas marked with *colored pixels*. The *color coding* indicates the number of macrocells each pixel belongs to, illustrating the overlapping level

implemented in the pre-upgrade MAGIC-I and used mainly for pulsar observations [8], so it will not be further discussed here.

3.2 Data Taking Procedure

All subsystems of MAGIC Telescopes are controlled and monitored from the Central Control (CC) program [9]. The CC defines standards for all observational procedures, and this section briefly describes what observation modes can be used and what types of data can be taken with the MAGIC Telescopes.

3.2.1 Source Pointing Modes

There are two observation settings used by MAGIC: the *tracking* (*ON/OFF*) mode and the *wobble* mode.

In *tracking* mode telescopes are pointing directly to the source, in such a way that the nominal position of the target is located at the center of each camera. Data recorded in this configuration are called *ON* data. In order to properly estimate the residual background in the *ON* sample, additional, dedicated observations of so-called *OFF* region(s) are required. *OFF* data are recorded from areas where no significant gamma-ray emission is expected, under the same circumstances as the *ON* observations (same zenith angle (*Zd*) range, level of background light, atmospheric conditions and so on).

Wobble, or the *false source tracking* mode [10], consists of observations where telescopes are alternatively tracking (at least) two directions on the sky, located with a slight offset with respect to the nominal position of the source (Fig. 3.6). For

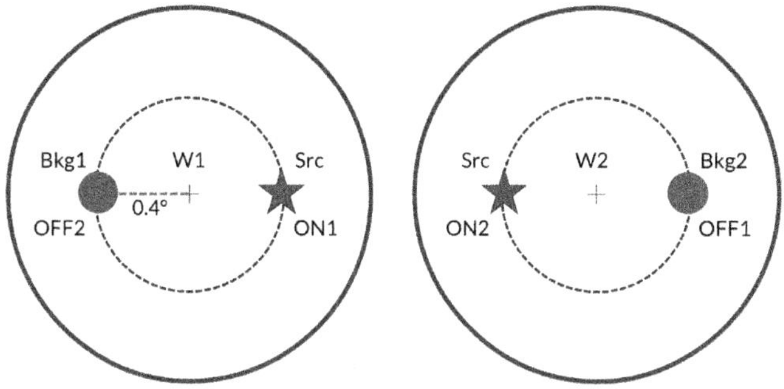

Fig. 3.6 Wobble observation mode: The residual background in the source region of W1 ($ON1$) is estimated from the background region of W2 ($OFF1$) and vice versa

MAGIC, the default offset is 0.4°, with wobble positions changing every 20 min to ensure uniform (azimuthal) coverage of the sky and avoid possible bias. The main advantage of this technique is that it does not require any additional *OFF* data, seen how the signal and the background are simultaneously measured. Background in every wobble is extracted from the opposite region to the source position in the camera, also called the anti-source. Such position is the source nominal position rotated by 180° around the camera center, and works as the equivalent to the *OFF* region in the tracking mode observations. Number of used background regions can be increased according to the user's needs, as long as these regions do not overlap with the signal region.

Still, the wobble mode observations have two main drawbacks: loss in the gamma-ray efficiency (by 15–20 %), due to the smaller effective trigger area around the source, and a possible bias introduced by the off-center source position, consequence of the camera inhomogeneities.

Furthermore, stereo observations originate an uneven acceptance along the FoV, referred to as the "stereo blob", caused by the broken azimuthal symmetry due to the relative orientations of the telescopes. The presence of the "blob" can be significantly irregular at the lowest energies, and it is most pronounced for observations carried out by only two instruments: the greater the number of telescopes in the system, the more symmetric the acceptance along the FoV.

3.2.2 Types of Data

Independently on the observation mode, MAGIC data are classified according to the kind of events that are recorded. Three kinds are always taken during the standard observations:

- *Pedestal run*: randomly triggered events (usually 2000 per run), digitized and recorded for measurement of the baseline value and evaluation of the effects the

NSB light and the readout electronics noise have. The contribution of both components is further extracted, pixel-by-pixel, in the calibration of the data (Sect. 3.3.1). Probability of recording an actual shower is negligible.

- *Calibration run*: events triggered by the calibration system of each telescope (Sect. 3.1.2), through generation of light pulses that mimic Cherenkov showers in terms of duration and wavelength. The calibration events are further applied in the calibration of the data (Sect. 3.3.1)—they are used to calculate the arrival time offsets as well as the conversion factor between the ADC counts and photoelectrons (ph.e.).
- *Data run*: events triggered by the cosmic showers. Data run also includes pedestal and calibration events, interleaved with the cosmic events at a rate of 25 Hz each. The interleaved events are further used to maintain the pedestal values and calibration constants updated during the sequential calibration of the data (Sect. 3.3.1).

Each telescope records, for every observed source, its own pedestal, calibration and data runs, under the common identifying run numbers assigned by the CC. In addition to these, in the case of MAGIC-II, a special kind of run is taken at the beginning of every night—the *Domino Calibration run* [6], used for correction of the non-linearity of the DRS2-based readout (Sect. 3.1.3.1). After the upgrade, this kind of run is no longer necessary.

3.3 Data Analysis

The main objectives in every analysis of MAGIC data are to distinguish between the gamma-ray and hadron events, to determine the primary energies of the gamma-ray photons and to precisely deduce their incoming directions.

The standard MAGIC analysis relies on programs and classes of the official MAGIC Analysis and Reconstruction Software (MARS) [11, 12], a dedicated software package written in C++ language and embedded in the framework of ROOT [13]. The philosophy of MARS is to create an analysis pipeline, where each program produces the input for the subsequent step. The final output is the list of events that is used for production of higher level results, like spectra and skymaps. The main links of the analysis chain are described in more detail in the following sections; this is their brief summary:

- *Data calibration*: for each pixel in the camera, charge (in ph.e.) and arrival time of Cherenkov pulses are determined.
- *Image cleaning and parametrization*: pixels that contain noise (and no signal) are removed from further analysis, after which parameters describing the shower image are computed.
- *Data selection*: quality control of the data is performed, based on acquisition rate stability, atmospheric conditions, values of the image parameters, etc.
- *Data merger*: data streams from both telescopes are combined together and the corresponding stereo parameters are calculated.

- *Event characterization*: for each event, its energy, arrival direction and likelihood to be a gamma-ray induced shower are estimated.
- *Calculation of the signal significance*: excess gamma-ray and background events are used to determine the significance of the observed gamma-ray signal.
- *Higher level products generation*: depending on the significance of the signal, differential and integral fluxes, the spectral energy distribution (SED) and light curves (in the case of the detection) or differential and integral upper limits to the flux (in the case of no detection) can be obtained. Skymaps can be created in either case.

The analysis chain relies on input from Monte Carlo (MC) simulations for background rejection and determination of the response function of the telescopes. Simulations include the EAS development, propagation of Cherenkov light through the atmosphere, response of the telescope mirrors and distribution of the photons on the camera plane, as well the response of the PMTs for a given NSB light and behavior of the readout [14, 15].

Additionally, if possible, analysis of the chosen source is always accompanied by a parallel analysis of a Crab Nebula data sample. Such sample should coincide with the principal source observations as much as possible, mainly in the atmospheric conditions and *Zd* range. The purpose of this complementary analysis is to examine the overall performance of the analysis chain. Crab Nebula data are used since this object is considered to be the standard candle of the VHE astronomy, given its intense and stable flux for the energy range of IACTs [16].

3.3.1 Data Calibration

This link of the analysis chain can be divided into several steps:

Conversion into ROOT Format MAGIC raw data consist of digitized pulses, recorded for every event and every pixel, with amplitudes expressed in arbitrary ADC units. Originally stored in binary form, these data are translated to ROOT format by means of the merging and preprocessing program (`merpp`). Aside from the format conversion, `merpp` is also responsible for injection of relevant reports from the telescope subsystems into data files. That way, all the information regarding the conditions and performance of the subsystems during data taking are easily accessible.

Signal Reconstruction For each pixel, the recorded signal of each event is sampled in 80, 0.5 ns-wide, ADC capacitors.[1] Every pulse is analyzed by the *signal extractor*, a routine that provides information on the arrival time of the signal as well as its amplitude and charge in ADC counts. Among various algorithms that can be used for the signal extraction [17], two are adopted in MAGIC analysis: the *cubic spline extractor* (for the pre-upgrade MAGIC-I data) and the *sliding window*

[1]The pre-upgrade DAQ of MAGIC-I was actually recording 50 out of 80 capacitors: the first and the last 15 capacitors only contained the “switching noise”.

extractor (for the pre-upgrade MAGIC-II and both telescopes post-upgrade). In the pre-upgrade MAGIC-II analysis, the signal extraction is always preceded by the correction of the data for the nonlinearity of the DRS2 chip (Sect. 3.1.3.1).

The *cubic spline extractor* works in the following way: after the pedestal subtraction, the ADC counts are interpolated by the cubic spline algorithm, and the maximum of the interpolation function is adopted as the signal amplitude. The signal charge is calculated as the integral of the interpolation function, computed in a fixed-width interval with limits dependent on the position of the maximum. Lastly, the location of the half maximum at the rising edge of the pulse gives the signal arrival time.

The *sliding window extractor* calculates the signal charge as the maximum integral content of 6 consecutive ADC capacitors, over a fixed time window and after the pedestal subtraction. The signal arrival time is defined as the average of the ADC capacitors time, weighted by the counts content in each of them.

ADC Count to Photoelectron Conversion Once the signal is reconstructed, its charge in ADC counts is translated into the equivalent number of ph.e., through the use of the *F-factor* method [18]. This approach makes the conversion based on the proportion ratio obtained from the calibration events. The *F-factor* method assumes an uniform photon detection efficiency over the entire camera, a number of incoming photons that follows the Poisson statistics, and the readout noise independent of the signal amplitude. If, on one hand, the distribution of the number of calibration events in each pixel has a mean N and RMS of $\sqrt{N}$, let the distribution describing the charge measured in ADC counts be of mean $\langle Q \rangle$ and RMS σ. The latter distribution is broader than the pure Poisson one, and their relative widths are related by the *F-factor* as:

$$F\frac{1}{\sqrt{N}} = \frac{\sigma}{\langle Q \rangle}. \tag{3.1}$$

The broadening of the measured distribution is a consequence of electron multiplication in the PMTs, and it has to be individually quantified for each PMT in the laboratory. In case of the pre-upgrade MAGIC-I, an averaged *F-factor* used for all PMTs was 1.14. For the pre-upgrade MAGIC-II and for both instruments after the upgrade, the *F-factor* is 1.095.

With known *F-factor* and measured $\langle Q \rangle$ and σ, Eq. (3.1) can be used to compute the mean number of ph.e. in calibration events:

$$C = \frac{N}{\langle Q \rangle} = F^2 \frac{\langle Q \rangle}{\sigma^2}, \tag{3.2}$$

where C is the conversion factor from ADC counts to ph.e. During the regular observations, response of the VCSELs varies and the values of the conversion factor change and thus have to be constantly updated. This is accomplished through the interleaved calibration events (Sect. 3.2.2).

The calibration of the MAGIC pre-upgrade data is performed by the program called `callisto` (calibrate light signals and time offsets), while after the upgrade this task is done by `sorcerer` (simple, outright raw calibration; easy, reliable extraction routines).

3.3.2 Image Cleaning and Parametrization

Cleaning After the calibration, the charge and arrival time information of each event in each pixel are used to identify noise generated by the NSB light and remove it from the shower image. This process is referred to as the *image cleaning*, and there are many algorithms that can fulfill this objective (for e.g., see [12, 19]). Performance of these algorithms, however, has a significant impact on the sensitivity of the analysis, especially at lowest energies, so the choice of the approach to be applied to the data has to be carefully considered. The image cleaning methods used in this work are the *absolute* cleaning and the *dynamic sum* cleaning. Both approaches first remove the pixels containing noise and then divide the remaining ones, that actually form the shower image, into two populations: the core and boundary pixels. The difference between two cleaning methods is the way the image core is defined.

In the absolute cleaning, a pixel is identified as of one or the other population if its charge exceeds a certain threshold of number of ph.e. (q_c for core, q_b for boundary pixels, with $q_c > q_b$) and if it is adjacent to at least one another pixel of the same population. Additionally, the individual arrival time of core pixel can not differ from the mean arrival time of the core of the image by more than Δt_c, while for boundary pixels the arrival time has to different by less than Δt_b from the arrival time of its core pixel neighbor. The standard values for the absolute image cleaning parameters are:

- MAGIC-I (pre-upgrade): $q_c = 6$ ph.e., $q_b = 3$ ph.e., $\Delta t_c = 4.5$ ns, $\Delta t_b = 1.5$ ns;
- MAGIC-II (pre-upgrade): $q_c = 9$ ph.e., $q_b = 4.5$ ph.e., $\Delta t_c = 4.5$ ns, $\Delta t_b = 1.5$ ns;
- MAGIC-II (post-upgrade): $q_c = 8$ ph.e., $q_b = 4$ ph.e., $\Delta t_c = 4.5$ ns, $\Delta t_b = 1.5$ ns.

Figure 3.7 shows the effects of the absolute image cleaning applied to the actual MAGIC-II events.

In the dynamical sum cleaning, the signals are clipped in amplitude and all possible combinations of 2NN, 3NN and 4NN multiplicities are summed up—if this sum is above a certain threshold, and within a sharp time interval, these pixels qualify as the image core. The clipping ensures that the PMT afterpulses (or strong NSB fluctuations) do not dominate the summed signal. After the 'sum', comes the 'dynamical' part of the process: the cleaning level of individual pixel is dynamically adapted for

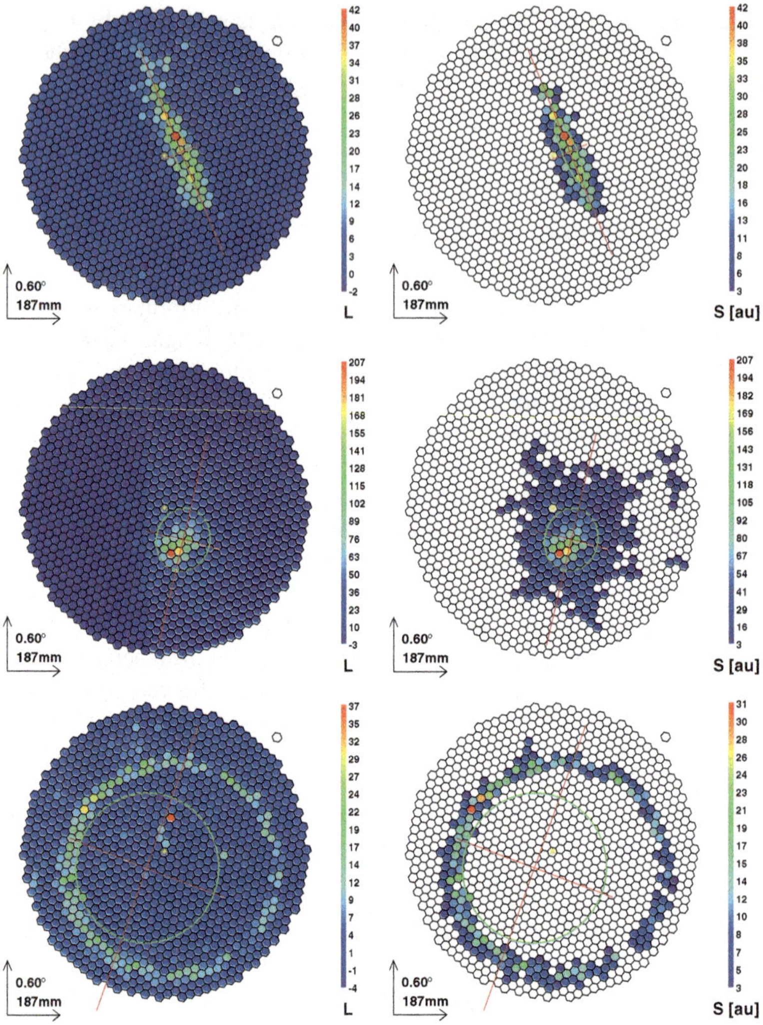

Fig. 3.7 Real events recorded with MAGIC-II, before (*left panels*) and after the image cleaning (*right panels*). *Top row* Gamma-like event. *Middle row* Hadron-like event. *Bottom row* Muon-like event

each event, depending on the total clipped sum of the charge in the shower core. Below a certain charge threshold, the scaling factor of the cleaning level q_c is 1, while for greater charges the scaling and the cleaning level increase. In addition, individual arrival time of core pixel can not differ more than a certain Δt_c from the mean arrival time of the shower core. Boundary pixels are defined as in the absolute cleaning, based on their charge and the arrival time differences with respect to the core neighbor pixels.

Since 2013, the dynamical sum cleaning is used as the default cleaning in the MAGIC analysis chain. For core pixels, the standard values of charge thresholds and time windows are 21.6, 24.6 and 25.2 ph.e. and 0.5, 0.7 and 1.1 ns for the 2NN, 3NN and 4NN groups; the cleaning level value of the individual core pixel, $q_c = 6$ ph.e. (same for both telescopes), is scaled for the clipped charge sum above a threshold of 750 ph.e. Other cleaning parameters are $q_b = 3.5$ ph.e., $\Delta t_c = 4.5$ ns, and $\Delta t_b = 1.5$ ns.

Parametrization The image cleaning is followed by reduction of each shower to a set of parameters quantifying its shape, orientation and timing. Based on their function, these parameters can be sorted as:

- **Hillas Parameters**, or the momenta of the 2-dimensional distribution of charge surviving the image cleaning [20]. The most relevant ones are:
  - *Size*: total charge (in ph.e.) of the full clean image. It is strongly correlated with the energy of the primary gamma-ray event.
  - *Width*: the RMS spread of the light along the minor axis of the image. It is a measure of the lateral development of the shower (Fig. 3.8).
  - *Length*: the RMS spread of the light along the major axis of the image. It is a measure of the longitudinal development of the shower (Fig. 3.8).
  - *Conc(N)*: fraction of the image charge concentrated in the *N* brightest pixels. It is an estimate of the compactness of the shower image, being usually larger for gamma-ray showers than for the hadronic ones.
- **Timing Parameters** exploit temporal properties of the shower:
  - *Time RMS*: the RMS spread of the arrival times of all the pixels that survived the image cleaning. It is a discriminator between the gamma-ray and hadronic showers, as their development times differ.
  - *Time gradient*: slope of the linear fit to the arrival time projection along the major axis line. It is a source-dependent parameter, adopting negative value if the arrival time increases when moving towards the location of the source in the camera, and positive otherwise.

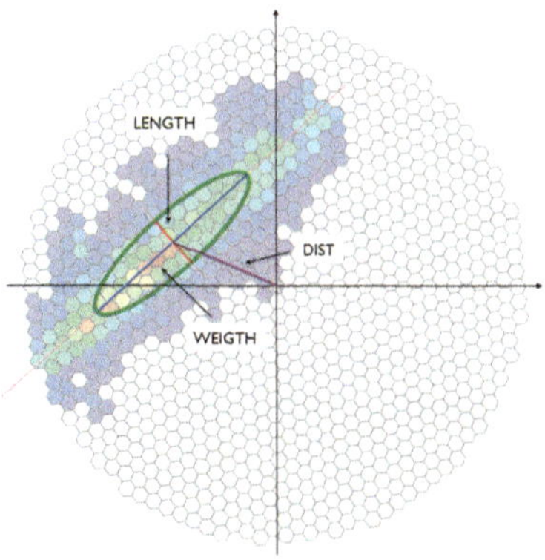

Fig. 3.8 Image of an actual shower in MAGIC-II camera after the cleaning, superimposed with the Hillas ellipse parametrization

- **Quality parameters** asses the image reconstruction accuracy:
  - *Leakage N*: fraction of the charge contained in the N outermost pixel rings of the camera. It serves as an estimator of the portion of the shower image that spills over the camera. Events with large *leakage N* are likely to undergo a bad reconstruction.
  - *Number of islands*: number of isolated groups of pixels that survive the image cleaning procedure. Gamma-ray showers generate a single island image, while the hadronic ones may produce several islands in a single event image (Fig. 3.7).

The image cleaning and parametrization of the single telescope data is performed by the standard analysis and reconstruction (`star`) program.

3.3.3 Data Selection

The purpose of data selection is to discard those data whose inadequate quality may jeopardize the sensitivity of the analysis. The main criteria in data selection are the atmospheric conditions and hardware performance during the observations.

Integrity of the hardware is constantly monitored by the subsystems of the telescopes, and possible malfunctions are reported in electronic runbooks of every observational night, or detected from the daily checks of the subsystems. Data taken with erroneous hardware, with problems that can not be corrected via software, are normally discarded from further analysis.

Bad atmospheric conditions are the main source of data degradation, as transparency of the atmosphere has significant impact on the propagation of Cherenkov light, and therefore on the resulting observations. A dedicated weather station, a lidar and a pyrometer operate at MAGIC site in parallel with observations and provide details on the current atmospheric circumstances. Nevertheless, the correlation between data quality and weather parameters is not conclusive, and there is not a sharply defined rule for data exclusion based on the atmospheric situation.

The most robust, systematic way for data quality check is monitoring of the events analysis rate r (after the image cleaning and above a certain *size* cut). It only depends on the observation *Zd* as:

$$r = r_0 \sqrt{\cos Zd}. \tag{3.3}$$

Thus, deviations from the expected value usually indicate some problem, and data whose rates differ for more than 20 % from the mean value are discarded.

Data quality can also be estimated from some other image parameter distributions, like *size*, *length* and *width*, seen how their values are expected to be constant during stable observations.

3.3.4 Data Merger and Stereo Parameter Reconstruction

Up to this point the analysis is applied to data of each telescope separately, but after the image parametrization and data selection, information about the same event recorded by individual telescopes are combined. This is performed by a program named `superstar`. Aside from data merger, `superstar` also calculates stereo parameters describing the 3-dimensional development of the shower [21], the most relevant of which are:

- *Shower axis*: information about the direction of the shower and its impact point on the ground. The projection of the arrival direction in the camera plane is estimated as the intersection of the major axes of the images. The impact point on the ground is determined from the intersection of the major axes of the images, taking into account the telescope positions (Fig. 3.9). The accuracy of estimating the shower axis depends on the relative positions of the telescopes and the shower: the more parallel the two images on the camera planes are, the larger the uncertainties.
- *Impact parameter*: the perpendicular distance between the shower axis and the telescope pointing axis (Fig. 3.9).
- *Shower maximum height*: estimate of the height at which the maximum development of the shower occurred. It is based on the shower axis characterization as the 3-dimensional geometrical location of the center of gravity (CoG) of the image. The shower maximum height is strongly correlated to the energy of the primary gamma-ray and is a powerful discriminator at low energies.
- *Cherenkov radius* and *Cherenkov photon density*: radius and density of the Cherenkov light pool on the ground. They are calculated assuming Cherenkov emission from a single electron at the shower maximum height and with energy equal to the critical energy, of 86 MeV.

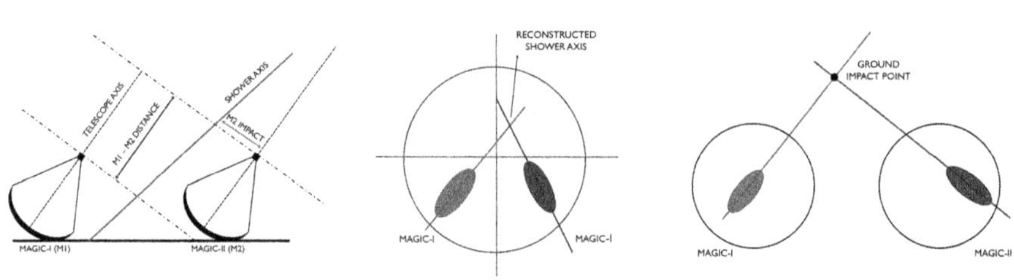

Fig. 3.9 Shower axis reconstruction in a stereoscopic view. *Left* Geometrical definition. *Center* Reconstruction of the shower direction as the intersection of the image major axes, once superimposed the images. *Right* Reconstruction of the shower impact point at ground

3.3.5 Event Characterization

The event characterization is aimed at determining the nature and properties of the primary particle that originated the shower. In particular, the energy and the arrival direction of the particle are of the greatest interest for the analysis. These values are estimated from the image parameters described in Sect. 3.3.2, following several different algorithms: the *Random Forest* (RF) type decision trees [22], parameterizations and *Look-Up Tables* (LUTs).

The characterization requires a MC dataset of simulated gamma-ray events, as it is not feasible to acquire pure gamma-ray sample from the observations. For the analysis purposes, the MC simulations are split into two subsets: one, called the *training* sample, used for preparation and tuning of the characterization algorithms, and the other, the *test* sample, used for evaluation of the performance of the trained algorithms. Both *train* and *test* samples are independent, thus avoiding biased results.

Before the actual characterization of the events, certain event-based quality cuts are applied. For instance, events with *size* lower than some minimum charge value are discarded; the same goes for events with too great *leakage*, too many *islands* and too few core pixels. Additionally, events that recorded sparks generated by electric discharges between the PMT cathodes and Winston cones are removed from the further analysis.

The estimation of the nature of the particle, its energy and its arrival direction are explained in more detail in the following. The program in charge of those tasks is called `melibea` (merge and link image parameter before energy analysis).

3.3.5.1 γ/Hadron Separation

IACT data are dominated by the background. Unwanted triggers are produced by fluctuations in the NSB light and showers originated by the cosmic rays, the so-called *hadronic showers* (Fig. 3.7, Sect. 2.3.4.2).

Hadronic showers are the most numerous population of recorded triggers: for observations of a typical VHE gamma-ray source, there are $\sim 10^4$ more hadronic than gamma-ray events. With a signal to noise ratio that small, sensitive measurements are only achievable if a highly effective hadron suppression is applied. In MAGIC analysis, this suppression is called *γ/hadron separation*, and there are several methods that provide it. The one used in this work (and by default in the MAGIC standard analysis) is the *RF method*, a multi-dimensional classification algorithm based on decision trees [23]. These trees are trained with a sample of MC simulated gamma-ray events and a sample of hadronic events from the actual observations. Trees grow in the space of image parameters, through dynamical evolution of the cuts on a randomly chosen set of variables. The final set of classification trees, defined as the mean classification from all trees, is referred to as *hadronness*—a likelihood that event is of hadronic origin. *Hadronness* takes values between 0 and 1: for the

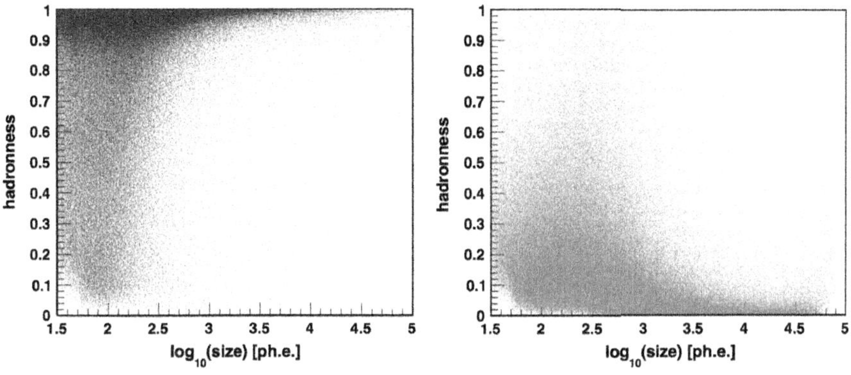

Fig. 3.10 Distribution of *hadronness* as a function of *size*, for a data sample (*left*) and for simulated MC gamma-ray events (*right*)

gamma-like events, *hadronness* is closer to 0, while for the hadronic ones it assumes greater values.

The γ/*hadron separation* for a given data set is performed by applying a cut in *hadronness* parameter. The resulting hadron suppression is about 90–99 % (improving with increasing *size* of the events).

Figure 3.10 illustrates the separation power of the *hadronness* parameter as a function of *size*.

3.3.5.2 Arrival Direction Estimation

As already mentioned in Sect. 3.3.4, direction of the shower can be calculated geometrically, but with some uncertainties dependent on the mutual positions of the telescopes. More accurate estimate of the arrival detection is achieved through the definition of a new parameter, called *disp*, that represents the angular distance between the image CoG and impact point in the camera. There are several methods that can be applied to estimate the value of *disp*. The one used in MAGIC stereo analysis, and in this work, is called *Disp RF*, and consists in introducing all image parameters that may influence the *disp* value in a dedicated RF algorithm [24]. The *Disp RF* is trained with a sample of simulated gamma-ray events of known source position, and it grows the corresponding decisional trees to evaluate the correlation between the *disp* and the input parameters. The value of *disp* is estimated for events from each telescope separately, and for the shower in each camera there are two possible reconstructed source positions along the image major axis (Fig. 3.11). When events are merged, the distances between all possible combinations of position pairs are calculated. The closest pair is chosen as the correct one, and the arrival direction is determined as the weighted average of such positions plus the crossing point of the main axes of the images.

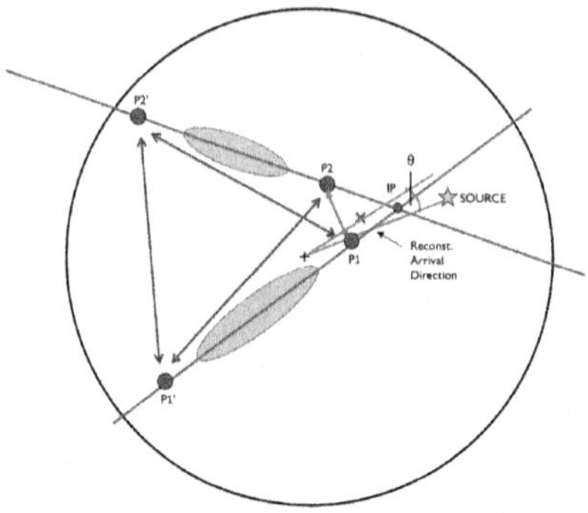

Fig. 3.11 Sketch of the stereo *Disp* algorithm. Each of the two superimposed images has two possible reconstructed source positions. The favored pair (P1-P2) is marked by the *green arrow*. The arrival direction (*red cross*) is reconstructed as the weighted average of P1, P2 and the point of intersection (IP) of two major axes

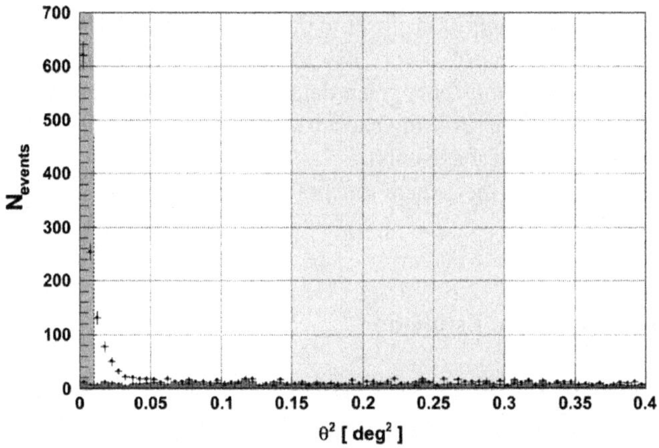

Fig. 3.12 Example of a signal detection from the Crab Nebula with a θ^2-plot. Excess gamma-ray events are given as points, while the background events are represented in *blue*. *Red shaded* area is the signal region, and the *green* one the normalization region. (Colour in online)

The use of *disp* parameter provides an unbiased way for estimation of the source position on the camera, as it does not require any previous assumption on its location. In addition, it allows for introduction of a new, powerful discriminator between the gamma-ray events coming from the observed source and the rest of the background: the θ^2 parameter. θ is defined as the angular distance between the reconstructed arrival direction of the event and the nominal source position on the camera plane (Fig. 3.11). For gamma-ray events from the source, the θ^2-distribution peaks toward zero values, while for hadronic and diffuse gamma-ray events, that are expected to fall isotropically, this distribution is flat (Fig. 3.12).

3.3.5.3 Energy Estimation

In the MAGIC standard analysis chain, the energy of each event is estimated through the use of LUTs. The LUTs are based on a simple model describing the distribution of the Cherenkov photons on the ground, by relying on the parameters *impact*, *Cherenkov radius* (r_c) and *Cherenkov density* (ρ_c), introduced in Sect. 3.3.4. LUTs are built for each telescope independently, by dividing a sample of simulated gamma-ray events in bins of *size* and in bins of *impact*/r_c ratio. Since the energy of the primary gamma-ray is proportional to *size*/ρ_c, each of the table bins contains the mean value of the distribution of $E' \times \rho_c/size$, where E' is the true energy, with an error given by the RMS of the distribution. The final energy estimation, E, is the average value obtained from both telescopes, weighted according to their errors, and corrected for the *Zd* dependence through the empirical factor $0.4 \times \cos(Zd)$.

3.3.6 Signal Estimation

After applying the *hadronness* cut, further background suppression can be achieved through the use of the θ^2 parameter. As already mentioned, the θ^2 distribution peaks close to zero values for gamma-ray events from the source, whereas for the background events, whose arrival directions are isotropic, the θ^2 distribution is flat. Consequently, by choosing a θ^2 cut below which the signal/background discrimination is good, geometrical space limited by that cut represents the signal (ON) region in the analysis (Fig. 3.12).

The events in the ON region are of both gamma-ray and hadronic origin. To estimate the residual background in the ON region, an independent data sample from the region with no expected gamma-ray signal (OFF) is used (see Sect. 3.2.1). The treatment of both background and signal regions, in sense of analysis methods and applied cuts, is the same. From the measured number of events in ON and OFF regions, N_{ON} and N_{OFF}, respectively, the number of excess events N_{ex} is calculated as:

$$N_{\mathrm{ex}} = N_{\mathrm{ON}} - \frac{N_{\mathrm{OFF}}}{\tau}, \tag{3.4}$$

where τ is the normalization factor between the ON and OFF samples, calculated as a fraction of N_{OFF} over N_{ON} events with high θ^2 values (Fig. 3.12).

For a given source, the detection significance (S) is computed by means of a statistical test, with null hypotheses assuming that the expected signal is no different from the background, i.e. all observed photons are from the background emission. The significance is calculated from the following, so-called Li&Ma expression [25]:

$$S = \sqrt{2}\left\{N_{\mathrm{ON}} \cdot \ln\left[(1+\tau)\left(\frac{N_{\mathrm{ON}}}{\tau N_{\mathrm{ON}} + N_{\mathrm{OFF}}}\right)\right] + \frac{N_{\mathrm{OFF}}}{\tau} \cdot \ln\left[(1+\tau)\left(\frac{N_{\mathrm{OFF}}}{\tau N_{\mathrm{ON}} + N_{\mathrm{OFF}}}\right)\right]\right\}^{1/2}. \tag{3.5}$$

As a convention, the detection of a source can be claimed if its significance level equals or surpasses $S = 5\sigma$ (referred to as 5σ detection[2]). This significance has to be corrected for the number of trials generated if different sets of cuts were used in the analysis, in order to avoid false detections caused by the possible background fluctuations.

The applied *hadronness* and θ^2 cuts are previously optimized on a reference Crab Nebula sample (Sect. 3.3), by maximizing its signal significance in the 2-dimensional parameter space.

Sensitivity The analysis can also be quantified in terms of its sensitivity, usually expressed as the minimum flux that would be measured above a certain energy threshold E_{th}, with 5σ significance in 50 h of observations, for a source whose spectrum is of the same shape as the one of the Crab Nebula. This *integral sensitivity* is calculated assuming a Gaussian approximation of the significance level (eq.(3.5)) and a normalization factor $\tau \to \infty$:

$$\Phi^{\text{min}}_{>E_{\text{th}}} = 5 \cdot \frac{\sqrt{N_{\text{OFF}}}}{N_{\text{ex}}} \sqrt{\frac{t_{\text{eff}}\ [\text{h}]}{50\ \text{h}}} \cdot \Phi^{Crab}_{>E_{\text{th}}}, \tag{3.6}$$

where $\Phi^{Crab}_{>E_{\text{th}}}$ is the Crab Nebula integral flux, with N_{ex} and N_{OFF} obtained from t_{eff} hours of Crab observations. The integral sensitivity is usually expressed as a total fraction of $\Phi^{Crab}_{>E_{\text{th}}}$, also called the Crab Nebula units (C.U.). Figure 3.13 shows the integral sensitivity of the MAGIC Telescopes.

The differential sensitivity is calculated in the same way, but with N_{ON} and N_{OFF} extracted in bins of estimated energy E. For a sufficiently fine binning, the spectral dependence vanishes.

3.3.7 *Higher Level Analysis Products*

Once the data have been selected, parametrized and reduced with the optimized analysis cuts, higher level results can be generated. For MAGIC, that means skymaps, light curves, integral and differential spectra, in the case of signal detection, or, otherwise, integral and differential upper limits. For the production of these results, however, it is necessary to know the response of the telescopes for gamma-rays of properties defined by the applied analysis cuts. Before describing in more detail higher level results of the MAGIC Telescopes, properties of its *response function* are briefly explained in the following section.

[2] A 5σ detection means that the probability of the detected signal resulting from background fluctuations is 3×10^{-7}.

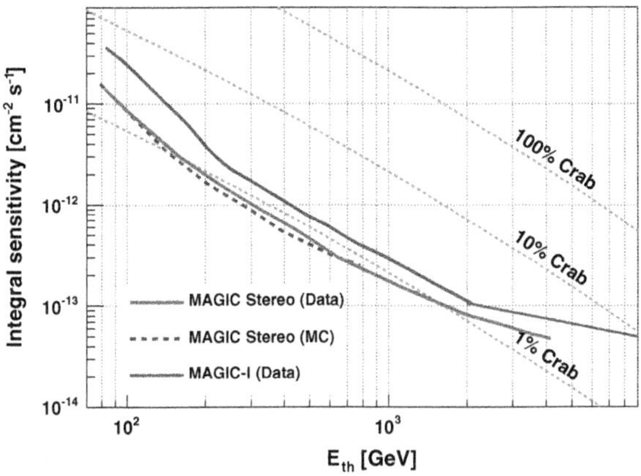

Fig. 3.13 The integral sensitivity of stereo observations as a function of threshold energy E_{th}, predicted from MC simulations (*dark red dashed line*) and measured from observations of the Crab Nebula (*solid red line*). *Blue line* represents the sensitivity achieved with MAGIC-I in standalone mode. For reference, different fractions of the Crab Nebula flux are represented as *gray dashed lines*. Taken from [26]. (Colour in online)

3.3.7.1 Response Function

The response function of an IACT is governed by its hardware design, reconstruction algorithms, selection criteria for quality of the events and for discrimination between gamma-rays and hadrons. It is computed by means of full MC simulations and for each analysis separately, as it depends on the particular cuts applied to the data as well as on the overall technical settings and performance of the instrument at the time of the observations.

The response function is typically represented as a combination of three different functions describing, for the given circumstances, the effective collection area, angular resolution and energy reconstruction parameters.

Effective Collection Area, A_{eff}, is the geometrical area around the telescope where gamma-ray shower produces a trigger (A_{sim}), folded with the gamma-ray efficiency ε_γ of all the cuts applied in the analysis. It is computed through MC simulations, and depends on the energy of the primary gamma-ray:

$$A_{\text{eff}}(E') = A_{\text{sim}}\varepsilon_\gamma(E') = A_{\text{sim}}\frac{N_{\gamma\text{AfterCuts}}(E')}{N_{\gamma\text{Total}}(E')}, \tag{3.7}$$

with ε_γ defined as the fraction of simulated MC gamma-ray events that survive all the analysis cuts ($N_{\gamma\text{AfterCuts}}$) and the total number of produced events ($N_{\gamma\text{Total}}$). The more constraining the cuts, the lower the ε_γ.

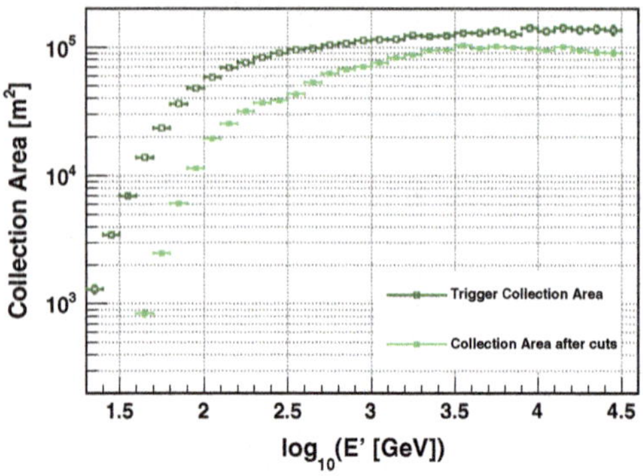

Fig. 3.14 MAGIC effective collection area for stereoscopic observations and at $Zd < 30°$. *Dark green open squares* show the collection area when only trigger efficiency is considered, while the *light green full squares* show the case when analysis cuts have been applied. Taken from [26]. (Colour in online)

A_{eff} also depends, for geometrical reasons, on the *Zd* of the observations: the projection on the ground of the light pool of showers with higher inclination is greater. In practice, A_{eff} is calculated in bins of *Zd*, and then combined as

$$A_{\text{eff}}(E') = \frac{\sum_i A_{\text{eff}\,i}\, t_{\text{eff}i}}{t_{\text{eff}}}, \tag{3.8}$$

where $t_{\text{eff}i}$ is the effective observation time of the i-th *Zd* bin.

Figure 3.14 shows an example of A_{eff} for the MAGIC stereoscopic observations: the probability to get a trigger increases very rapidly at low energies until all the showers in the light pool generate enough light to induce triggers, after which A_{eff} remains almost constant.

Angular Resolution is also referred to as the PSF (point spread function) of the telescope. It describes the capability of the instrument to reconstruct the gamma-ray incident direction, and it is defined as the standard deviation of the 2-dimensional Gaussian fit to the sky distribution of a point-like source. The PSF corresponds to the radius containing 39 % of the gamma-ray events from the source. Figure 3.15 shows the stereo angular resolution of MAGIC: for 300 GeV it is as good as 0.07°, and even better at higher energies. For the sake of completeness, also shown is the 68 % containment radius.

Energy Reconstruction is described by the *energy resolution* and *energy bias*. The energy resolution σ is defined as the width of a Gaussian fit to the $(E - E')/E'$ distribution, where E' and E are the true and estimated energies, respectively. The mean of the fit is the relative energy bias μ. The energy reconstruction is as good as those two parameters are close to zero. The values of σ and μ for MAGIC stereo observations are shown on Fig. 3.16.

Additionally, E_{th} of the analysis is conventionally defined as the peak of the energy distribution from a simulated gamma-ray *test* sample, once the background rejection and analysis cuts are applied. E_{th} strongly depends on the minimum cut in the *size* image parameter.

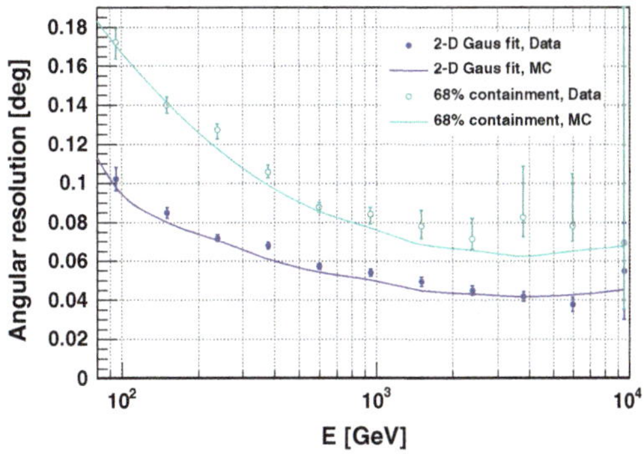

Fig. 3.15 MAGIC angular resolution (*violet*) and 68 % containment radius (*cyan*), as computed from a MC-simulated point-like source (*lines*) as well as from a Crab Nebula sample (*circles*). Taken from [26]

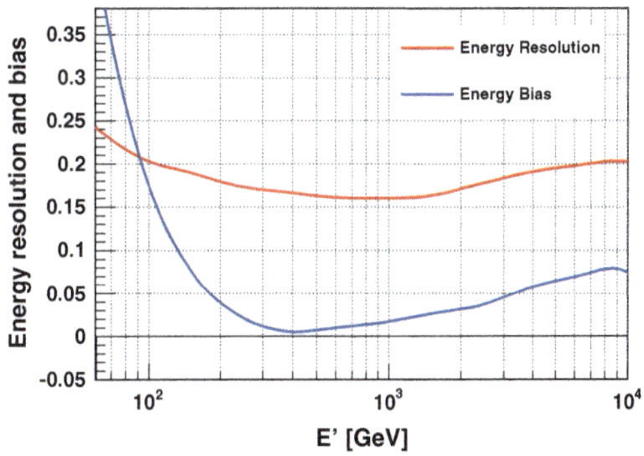

Fig. 3.16 MAGIC energy reconstruction parameters: *Red line* represents the energy resolution, the *blue* one the energy bias. Taken from [26]

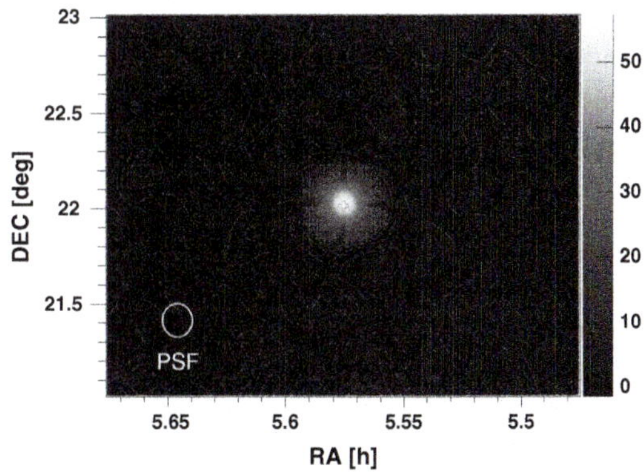

Fig. 3.17 Example of the skymap obtained from the Crab Nebula data sample

3.3.7.2 Skymap

Skymap refers to a 2-dimensional histogram that contains arrival directions, in sky coordinates, of all gamma-ray events that remain after the analysis cuts and after the subtraction of the expected background, smoothed with a Gaussian of width equal to the angular resolution at a given energy (Fig. 3.17).

Skymap production relies on unbiased reconstruction of the arrival time provided by the *disp* parameter (Sect. 3.3.5.2), as well as on the correct modeling of the background, which often is not a straightforward task. The camera acceptance in general is not homogeneous, and this irregularity is additionally boosted by factors like field stars, malfunctioning pixels and trigger inefficiencies. Furthermore, the dependence that the sensitivity of the instrument has on the *Zd* must be taken into account. For different observation modes (Sect. 3.2.1), background is estimated differently. For the tracking mode, background is built from the OFF sample, assuming an isotropic arrival direction of the events. In the wobble mode, the camera inhomogeneities are somewhat smoothed: for each wobble position, the camera is divided into two halves, one containing the source position and the other background events. From there, background is modeled only from the events whose arrival directions do not lie in the source part of the camera [27].

Skymaps of well-known sources are also used to check the mispointing of the instrument.

3.3.7.3 Spectra

The differential gamma-ray spectrum is defined as the total number of photons coming from the source and reaching the observer, in unit of energy, area and time:

$$\frac{d\Phi}{dE} = \phi(E) = \frac{dN_\gamma}{dt_{\text{eff}}\, dA_{\text{eff}}(E)\, dE}, \tag{3.9}$$

where t_{eff} is the *effective observation time*, i.e. time during which telescope has been recording events, corrected for the dead time of the readout system electronics.

From the computational point of view, the differential treatment of A_{eff} in Eq. (3.9) translates into fine energy binning. As for the dt_{eff} term, it is replaced by a single temporal bin, of duration t_{eff}, encompassing all the observation and within which the emission from the source is considered steady.

Spectrum Unfolding The spectrum obtained by the above described procedure is affected by distortions in the number of events of each bin due to the finite energy resolution and bias of the detector. In particular, the spectrum is calculated as a function of estimated energy E (which has been computed as a function of image parameters); but, for the spectrum to have a physical meaning, it has to be given as a function of the true energy E'. Therefore, the spectrum has to be *unfolded*, by transforming the measured distribution $g(E)$ to the true energy distribution $f(E')$:

$$g(E) = \int M(E', E) f(E') dE', \tag{3.10}$$

where $M(E', E)$ is the migration matrix, describing the response of the detector; it can be obtained directly from MC simulations, by applying to the simulated gamma-ray events the same energy estimation as for the real data. There are well-known problems associated to the direct solution for $f(E')$, defined as that obtained by the exact inversion of Eq. (3.10). It can be shown that the exact solution is an estimator of the true spectrum, with zero bias but diverging variance. The problem is solved in practice by constructing physically meaningful estimators with a controlled amount of bias and much reduced variance. This technique is generally known as spectrum unfolding, and the differences among the various approaches come from the various methods to measure and control the bias through regularization functions. Some of the of unfolding approaches implemented in the MAGIC analysis chain include forward unfolding [28], Tikhonov [29], Bertero [30] and Schmelling [31] methods.

3.3.7.4 Light Curve

The variability of a source is represented by means of a light curve—the flux of the source presented in different time bins. All points of the light curve cover the same energy interval, $[E_{\min} - E_{\max}]$. For bin i, spanning over the time interval $[t_m^i - t_n^i]$, the corresponding integral flux is calculated as:

$$\Phi^i(E) = \int_{t_m^i}^{t_n^i} \int_{E_{\min}}^{E_{\max}} \phi(E) dE dt. \tag{3.11}$$

This way of computing a light curve point only applies if the source is detected with enough significance in the given energy range and time interval. Otherwise, the calculated value is not the integral flux but an upper limit to the integral flux.

3.3.7.5 Upper Limits

Whenever no significant detection can be claimed, upper limits to the differential or integral flux of the source can be derived, with a certain confidence level (c.l.).

For the upper limit calculations, the necessary input includes the measured number of events from the source and background regions (N_{ON} and N_{OFF}, respectively), as well as the assumption on the energy differential spectrum, in simple form given as $\phi(E) = K \cdot S(E)$, where K is the normalization constant. Additionally, the possible emission is presumed to be constant within the considered time interval.

In MAGIC analysis, the standard way to calculate upper limits is by estimating the maximum number of excess events ($N_{\text{ex}}^{\text{UL}}$), present in the *ON* region within the chosen c.l., through the so-called Rolke method [32], also referred to throughout this work as the conventional approach (Sect. 4.1.1; Eq. (4.1)). In addition, the spectrum is approximated by a simple power law of spectral slope α (= −2.6 for the Crab-like spectrum). From there, the integral flux upper limit can be expressed through the normalization constant K as:

$$K^{\mathrm{UL}} < \frac{N_{\mathrm{ex}}^{\mathrm{UL}}}{t_{\mathrm{eff}} \int_{E_{\min}}^{E_{\max}} S(E) A_{\mathrm{eff}}(E) dE}. \tag{3.12}$$

As the conventional method is a fully frequentist approach, uncertainties in the background estimation are treated as nuisance parameters. In MAGIC standard analysis chain, these uncertainties are set to 30 %, based on the systematic errors of the analysis method whose efficiency is taken to be constant and equal to 1. The c.l. is by default chosen to be 95 %.

However, the conventional method does not provide the optimal sensitivity when $\phi(E)$ has some characteristic features; as a part of this work, a dedicated approach has been developed to maximize the analysis sensitivity for such spectra. Detailed explanation is provided in Chap. 4.

3.3.8 Systematic Uncertainties

While for the weak sources the main uncertainty in the analysis is caused by low statistics of excess gamma-ray events, for strong sources the systematic errors become the dominant ones. The principal factors contributing to the systematic uncertainties can be grouped as:

- *Sources of systematic uncertainties on energy scale*: atmospheric transmission, insufficiently good knowledge on the current state of the hardware components (mirror reflectivity, QE of the PMTs, photon losses on the camera entrance (Sect. 3.1.1)) and effectiveness of analysis methods (signal extraction, calculation of the *F-factor* (Sect. 3.3.1), etc.) These uncertainties affect the energy scale by $\sim$17 % at low energies ($\leq$ 100 GeV) and $\sim$15 % at medium ones ($\geq$ 300 GeV).
- *Sources of systematic uncertainty on the flux normalization*: discrepancies between the MC simulations and the data, background subtraction, camera inhomogeneities and faulty pixels, mispointing, NSB, dead time of the readout. For low and medium energies, these uncertainties affect the flux normalization by $\sim$19 % and $\sim$11 %, respectively.
- *Sources of systematic uncertainty on spectral slope*: different efficiencies of various approaches applied in the analysis (different unfolding methods (Sect. 3.3.7.3), different efficiencies of the selection cuts (Sect. 3.3.3) and different *Zd* ranges). Also, it includes the uncertainties caused by the non-linearity of the pre-upgrade MAGIC-II readout (Sect. 3.1.3.1). The slope of the differential spectrum is affected by the uncertainty of $\pm$0.15.

3.4 Accessibility of the Analysis Results

Once the analysis has been completed, and its outcome cross-checked by independent member(s) of the MAGIC Collaboration, the final results are made public. They can be accessed either in the printed form, through scientific journals, or in a digital form, as FITS files.

This section describes in more detail the latter way of presenting MAGIC results, as the Author of this work has been responsible for the development of the tools and maintenance of the database of the MAGIC results in FITS form and of a system for their accessibility through the Virtual Observatory (VO).

3.4.1 FITS File Format

FITS stands for "Flexible Image Transport System" [33], and it is a standard file format used to store, transmit and manipulate scientific data sets. The format is designed as self-defining and adaptable to changing needs that may arise from different applications. One of the FITS main features is that it can carry any number of n-dimensional data arrays (like 1-dimensional spectra, 2-dimensional images, 3-dimensional data cubes,...). Furthermore, FITS stores all the metadata information associated with such matrices in human-readable headers.

FITS structure is based on so-called header and data units (HDUs), that contain the metadata information written as ASCII text, followed by an integer number of binary data records. The FITS file itself may be composed out of any number of HDUs, out of which the first one is the primary header (PHDU) and the rest are the extensions. HDUs are defined based on their content and can be of the type IMAGE or BINTABLE.

The ASCII cards contained in the headers are 80 character fixed-length strings that carry keyword/value pairs. The keywords are used to provide information related to the data, such as its size, origin, coordinates, analysis details or any other information the creator might want to include. Some keywords are defined by the FITS standards (reserved keywords), while others may be defined according to the needs of the user.

3.4.2 FITS Format for MAGIC Data

First standardized in 1981, FITS is the most generally used format in astronomy today. However, within the VHE astronomical community (and especially among the international collaborations operating the IACTs), it is not as popular. The reason for this probably lies in the fact that, historically, this field has been more close to Particle Physics and its way of presenting the data. Nevertheless, in the last few years this trend has been gradually changing, particularly in the light of projected construction of the next generation IACT—the CTA. CTA is intended to operate as a "standard" astronomical observatory and to provide public access to significant portion of its data in the FITS form (Sect. 6.1, [34]). In the meantime, the IACTs of the current generation have been working, each one in their own way, on releasing their results in the FITS format.

The MAGIC Collaboration started its public FITS Database in 2008 [35]. It is based on a principle that for each scientific work published by the Collaboration in a refereed international journal, there is one associated FITS file. The information contained in such files are all the higher level results from the printed publication. The exception are the publications of some multi-wavelength or joint IACTs campaigns: while results from other observatories are presented in the article, MAGIC does not publicly distribute such information in the corresponding FITS files. Also, some particular, telescope-specific findings that, although present in the publication, are considered to be of no relevance for an external analyzer, are excluded from the FITS file as well.

According to the MAGIC Standard [36], structure of the FITS file consists of the compulsory PHDU, followed by any number of extensions. The extension HDUs may contain skymaps (Sect. 3.3.7.2), integral and differential spectra (Sect. 3.3.7.3), integral and differential upper limits (Sect. 3.3.7.5), SEDs, light curves (Sect. 3.3.7.4), alpha-[3] and θ^2-plots (Sect. 3.3.5.3), and even other kinds of 1- or 2-dimensional distributions with user-defined magnitudes presented on the x and y axes.

Keywords that can be used in the PHDU and HDUs are defined in accordance with the MAGIC observations and analyses. Apart from some required and reserved words, some new, MAGIC (or IACT) specific variables have been defined as well. For instance, the PHDU contains general information about the FITS file itself: when and by whom it was created, to which publication it refers to, and what is its current version. Furthermore, it gives details about the observations (when did they take place, with which instrument and at which *Zd* range), about the source (its name, coordinates, periodicity and phase,...) and some global characteristics of the performed analysis (considered energy range, cuts, the assumed c.l. for exclusion limits, t_{eff}, gamma-ray efficiency, etc.). The PHDU only contains information that is common for the entire file; everything that applies only to a certain higher level product is written in the header of the corresponding extension instead. Furthermore, aside from these general keywords, there are some that are specific for the given extension (like whether the spectrum is differential or integral, if the skymap been smeared with a Gaussian function and of what width...).

For the creation of MAGIC FITS files, a tool called `MFits` is used. `MFits` is written as a class of ROOT [13], and it relies on the public libraries `CCfits` [37] and `cfitsio` [38]. Aside from creation of FITS files for the Database, `MFits` can also process the lower level results of MAGIC analysis, as well as make the two-way conversions between the FITS and ROOT format. At the moment, `MFits` is only available to the members of the MAGIC Collaboration.

[3]Alpha-plot may be considered the mono-observations-equivalent to the θ^2-plot. As it is not used in this work, further details are omitted. More information can be found in [9].

3.4.3 MAGIC Data at the Virtual Observatory

Over the years, the amount (and size) of data gathered by astronomical facilities has been increasing, together with variety of scientific tools needed for the interpretation and analysis of different datasets. It was becoming evermore difficult for the astronomers to cope with such an overwhelming abundance. The issue was addressed through the creation of the Virtual Observatory (VO)—a collection of interoperating data archives and software tools which utilizes the internet to form a scientific research environment in which astronomical research programs can be conducted [39].

The core components of the VO are finding what data are available and getting access to them through simple requests. For this concepts to function, data have to be published in a VO-compliant form—that is, according to VO standards that make the data readable to many of the commonly used scientific tools. VO standards are set by the International Virtual Observatory Alliance (IVOA) [40], an organization that defines the protocols and coordinates the efforts of different VOs. At the moment, the VO standards are not adapted for the needs of IACTs: for instance, some of the required fields in the VO searches are not optimized for Cherenkov observatories, and some data types (like θ^2-plots) are not supported. However, these standards are constantly evolving in accordance with the needs of the astronomical community.

MAGIC public data are available in a VO-compliant form, either through the registry of the European Virtual Observatory [39] or directly from the MAGIC VO server (Fig. 3.18, [41]). For the time being, only spectrum and light curve protocols

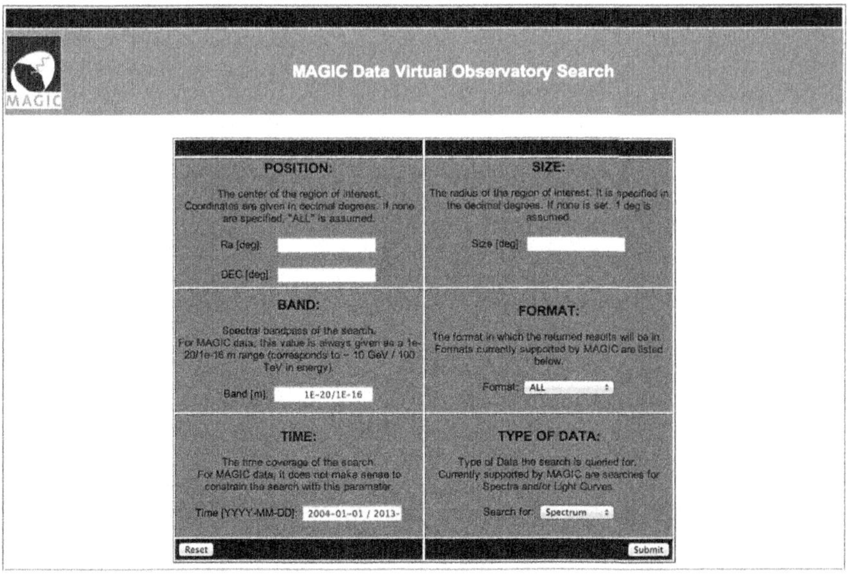

Fig. 3.18 Screenshot of the MAGIC VO interface, [41]

are supported, with the one for the skymaps being currently implemented. Data are provided to the user in FITS format, with one file created for each search result. Tools that manage VO requests are also incorporated in the `MFits` class.

References

1. The MAGIC Collaboration, MAGIC Telescopes home page, http://magic.mppmu.mpg.de. Dec 2012
2. J. Sitarek et al., Physics performance of the upgraded MAGIC Telescopes obtained with Crab Nebula data. To appear in the *Proceedings of the 33rd International Cosmic Ray Conference*, Rio de Janeiro, Brasil (2013)
3. T. Bretz et al., The drive system of the Major Atmospheric Gamma-ray Imaging Cherenkov Telescope. Astropart. Phys. **31**, 92–101 (2009). arXiv:0810.4593
4. M. Doro, Novel reflective elements and indirect dark matter searches for MAGIC II and future IACTs, Ph.D. thesis, Universitá degli Studi di Padova, Italy, 2009
5. F. Goebel et al., Upgrade of the MAGIC Telescope with a multiplexed Fiber-Optic 2G Samples/s FADC Data Acquisition system, in *Proceedings of the 30th International Cosmic Ray Conference*, Merida, Mexico (2007), arXiv:0709.2363
6. J. Aleksić, Characterization of the MAGIC II Data Acquisition System. M.S. thesis, Universitat Autònoma de Barcelona, Spain, 2009
7. D. Tescaro, The Upgraded Readout System of the MAGIC Telescopes, in *Proceedings of the IEEE Nuclear Science Symposium*, Anaheim, California (2012)
8. M. Rissi et al., A new sum trigger to provide a lower energy threshold for the MAGIC Telescope. IEEE Trans. Nucl. Sci. **56**, 3840–3843 (2009)
9. R. Zanin, J. Cortina, The central control of the MAGIC Telescopes, in *Proceedings of the 31st International Cosmic Ray Conference*, Łódź, Poland (2009), arXiv:0907.0946
10. V.P. Fomin et al., New methods for atmospheric Cherenkov imaging for gamma-ray astronomy 1: the false source method. Astropart. Phys. **2**, 137–150 (1994)
11. A. Moralejo et al., MARS, the MAGIC analysis and reconstruction software, in *Proceedings of the 31st International Cosmic Ray Conference*, Łódź, Poland (2009), arXiv:0907.0943
12. S. Lombardi et al., Advanced stereoscopic gamma-ray shower analysis with the MAGIC Telescopes, in *Proceedings of the 32nd International Cosmic Ray Conference*, Beijing, China (2011), arXiv:1109.6195
13. The ROOT Team, A Data Analysis Framework, http://root.cern.ch/drupal. Dec 2012
14. P. Majumdar et al., Monte Carlo simulation for the MAGIC Telescope, in *Proceedings of the 29th International Cosmic Ray Conference*, Pune, India (2005)
15. E. Carmona et al., Monte Carlo simulation for the MAGIC-II system, in *Proceedings of the 30th International Cosmic Ray Conference*, Merida, Mexico (2007), arXiv:0709.2959
16. R. Zanin, Observations of the Crab Nebula pulsar wind nebula and microquasar candidates with MAGIC. Ph.D. thesis, Univeristat Autònoma de Barcelona, Spain, 2011
17. J. Albert et al., FADC signal reconstruction for the MAGIC Telescope. Nucl. Instrum. Meth. A **594**, 407–419 (2008). astro-ph/0612385
18. R. Mirzoyan, On the calibration accuracy of light sensors in atmospheric Cherenkov fluorescence and neutrino experiments, in *Proceedings of the 25th International Cosmic Ray Conference*, Durban, South Africa (1997)
19. E. Aliu et al., Improving the performance of the single-dish Cherenkov telescope MAGIC through the use of signal timing. Astropart. Phys. **30**, 293–305 (2009). arXiv:0810.3568
20. A.M. Hillas, Cherenkov light images of EAS produced by primary gamma rays and by nuclei, in *Proceedings of the 19th International Cosmic Ray Conference*, La Jola, USA (1985)
21. A. Kohnle et al., Stereoscopic imaging of air showers with the first two HEGRA Cherenkov Telescopes. Astropart. Phys. **5**, 119–131 (1996)

22. L. Breiman, Random Forests. Mach. Learn. **45**, 5–32 (2001)
23. J. Albert et al., Implementation of the random forest method for the imaging atmospheric Cherenkov telescope MAGIC. Nucl. Instrum. Meth. A **588**, 424–432 (2008). arXiv:0709.3719
24. J. Aleksić et al., Search for an extended VHE -ray emission from Mrk 421 and Mrk 501 with the MAGIC Telescope, A&A 524, A77 (2010), arXiv:1004.1093
25. T.P. Li, Y.Q. Ma, Analysis methods for results in gamma-ray astronomy. Astrophys. J. **272**, 317–324 (1983)
26. J. Aleksić et al., Performance of the MAGIC stereo system obtained with the Crab Nebula data. Astropart. Phys. **35**, 435-448 (2012), arXiv:1108.1477
27. M. Errando, Discovery of very high energy gamma-ray emission from 3c 279 and 3c 66A/B with the MAGIC Telescope. Ph.D. thesis, Univeristat Autònama de Barcelona, Spain, 2009
28. W. Wittek, *Unfolding (revised), MAGIC internal note TDAS05-05*, Max-Planck-Institut for Physics, Munich, Germany (2009)
29. A.N. Tikhonov, V.Y. Arsenin, *Solutions of ill posed problems*, vol. H, 1st edn. (Winston, New York, 1977). ISBN 978-0470991244
30. M. Bertero, Linear inverse and ill-posed problems. in *Advances in Electronics and Electron Physics*, vol. 75, ed. by P.W. Hawkes (Academic Press INC, San Diego, 1989), pp. 1–120. ISBN: 978-0120146758
31. M. Schmelling, Numerische Methoden der Datenanalyse, *BlockkursWinter Semester 1997/1998* (MPI-K Heidelberg, Germany, 1998)
32. W.A. Rolke, A.M. López, J. Conrad, Limits and confidence intervals in the presence of nuisance parameters. Nucl. Instrum. Meth. A **551**, 493-503 (2005), arXiv:physics/0403059
33. R.J. Hanisch et al., Definition of the Flexible Image Transport System (FITS). A&A **376**, 359–380 (2001)
34. The CTA Consortium, CTA Data and e-Infrastructure, https://www.cta-observatory.org/?q=node/178. Dec 2012
35. The MAGIC Collaboration, MAGIC Datacenter @ PIC—FITS database, http://magic.pic.es/pub/_ts. Dec 2012
36. J. Aleksić, J. Rico, *FITS format for MAGIC data, MAGIC internal note TDAS 09-05*, Institut de Física d'Altes Energies, Barcelona, Spain (2009)
37. NASA Goddard Space Flight Center, CCfits: Wrappers for the cfitsio library, http://heasarc.gsfc.nasa.gov/_tsio/CC_ts Dec 2012
38. NASA Goddard Space Flight Center, FITSIO home page, http://heasarc.gsfc.nasa.gov/_tsio. Dec 2012
39. EURO-VO, European Virtual Observatory, http://www.euro-vo.org/pub/general/qa.html. Dec 2012
40. IVOA, International Virtual Observatory Alliance, http://www.ivoa.net. Dec 2012
41. The MAGIC Collaboration, MAGIC Data Virtual Observatory Search, http://vobs.magic.pic.es. Dec 2012

Chapter 4
Full Likelihood Method

Cherenkov telescopes are observatories that deal with great variety of scientific objectives—from detection and study of galactic and extragalactic gamma-ray sources to probing some of the most intriguing questions of the fundamental physics and cosmology. However, the duty cycles of these instruments are limited, and not all of the issues can be properly covered. The preference is usually given to the astrophysical objects of conventional origin, with spectral distributions nicely described, in the majority of cases, by a simple power law. As a consequence, standard analysis tools and methods are adapted for such signals, at the expense of sources whose emissions are predicted to contain some distinctive spectral features.

This chapter is devoted to the introduction of an alternative analysis method, *full likelihood*, optimized for the recognition of spectral features in IACT observations. In the upcoming sections, concepts and characterization of the full likelihood will be presented, as well as a comparison of its performance with respect to the standard, *conventional* approach, currently deployed in the IACT analyses.

4.1 The Method

Observations with IACTs are dominated by the background. For each gamma-ray photon from the source (*ON*) region one may expect thousands of unwanted cosmic ray protons, the majority of which are later removed through the analysis cuts (Sect. 3.3.5.1). The number of background events that remain in the data sample is estimated with high precision from the synchronous, or very similar (to the source) observations of control (*OFF*) regions from which no gamma-ray emission is expected (Sect. 3.2.1).

This section describes how the acquired data are analyzed by the conventional and by the full likelihood methods.

J. Aleksić, *Optimized Dark Matter Searches in Deep Observations of Segue 1 with MAGIC*, Springer Theses, DOI 10.1007/978-3-319-23123-5_4

4.1.1 Conventional Analysis Approach

In the standard analysis chain of IACTs, the existence of a source is established by a mere comparison of the integrated number of events detected from the *ON* region (N_{ON}) with the integrated number of events from the *OFF* region(s) (N_{OFF}) (Sect. 3.3.6, Eq. (3.5)). Both N_{ON} and N_{OFF} are random variables that obey Poisson statistics; therefore, the actual number of gamma-ray (g) and background events (b) present in the *ON* region can be estimated through maximization of the following likelihood function [1]:

$$\mathscr{L}(g, b|N_{\mathrm{ON}}, N_{\mathrm{OFF}}) = \frac{(g+b)^{N_{\mathrm{ON}}}}{N_{\mathrm{ON}}!}e^{-(g+b)} \times \frac{(\tau b)^{N_{\mathrm{OFF}}}}{N_{\mathrm{OFF}}!}e^{-\tau b}, \tag{4.1}$$

with τ denoting the normalization between the *ON* and *OFF* regions (e.g. ratio of their associated observation times).

This approach is what is currently used in the standard analysis of the IACTs (Sect. 3.3.7.5) and which is referred to, in this work, as the *conventional* approach. Whilst acceptable for sources of astrophysical origin, this method makes no distinction of the potential features present in the gamma-ray spectrum, and as such, it is suboptimal for studies where such signatures play a significant role (e.g. dark matter searches, Sect. 2.3.2.3).

4.1.2 Full Likelihood Method

As an alternative to the conventional approach, the *full likelihood* method is based on making an a priori assumption on the expected spectral shape (which is fixed for the chosen signal model), and including it in the maximum likelihood analysis. That way, the spectral information of the signal events if completely exploited, and the achieved sensitivity of the analysis increased with respect to that of the conventional method.

The full likelihood function has, for a given signal model M with parameters $\boldsymbol{\theta}$, the following form:

$$\mathscr{L}(N_{\mathrm{EST}}, M(\boldsymbol{\theta})|N_{\mathrm{OBS}}, E_1, \ldots, E_{N_{\mathrm{OBS}}}) = \frac{{N_{\mathrm{EST}}}^{N_{\mathrm{OBS}}}}{N_{\mathrm{OBS}}!}e^{-N_{\mathrm{EST}}} \times \prod_{i=1}^{N_{\mathrm{OBS}}} \mathscr{P}(E_i; M(\boldsymbol{\theta})), \tag{4.2}$$

with N_{OBS} ($= N_{\mathrm{ON}} + N_{\mathrm{OFF}}$) and N_{EST} denoting the total number of observed and estimated events, respectively, in *ON* and *OFF* regions.

$\mathscr{P}(E_i; M(\boldsymbol{\theta}))$ is the value of the probability density function (PDF) of the event i with measured energy E_i. In general, $\mathscr{P}$ can also depend on the measured direction of the photon. However, for the purposes of this work, contribution of that parameter is integrated out in the analysis (for more details, see Sect. 4.2.1.1). Therefore, the

PDF is defined as a function of measured energy only:

$$\mathscr{P}(E;M(\boldsymbol{\theta})) = \frac{P(E;M(\boldsymbol{\theta}))}{\int_{E_{\min}}^{E_{\max}} P(E;M(\boldsymbol{\theta}))dE}, \tag{4.3}$$

where $E_{\min}$ and $E_{\max}$ are the lower and upper limits of the considered energy range. $P(E;M(\boldsymbol{\theta}))$ represents the differential rate of the events, such that:

$$P(E;M(\boldsymbol{\theta})) = \begin{cases} P_{\mathrm{OFF}}(E_i), & i \in OFF \\ P_{\mathrm{ON}}(E_i;M(\boldsymbol{\theta})), & i \in ON \end{cases}, \tag{4.4}$$

with $P_{\mathrm{OFF}}(E)$ and $P_{\mathrm{ON}}(E;M(\boldsymbol{\theta}))$ being the expected differential rates from the *OFF* and *ON* regions, respectively:

$$P_{\mathrm{OFF}}(E) = \tau \int_0^\infty \frac{d\Phi_B}{dE'} R_B(E;E')dE' \tag{4.5}$$

and

$$P_{\mathrm{ON}}(E;M(\boldsymbol{\theta})) = \int_0^\infty \frac{d\Phi_B}{dE'} R_B(E;E')dE' + \int_0^\infty \frac{d\Phi_G(M(\boldsymbol{\theta}))}{dE'} R_G(E;E')dE'. \tag{4.6}$$

True energy is denoted with E'. $d\Phi_B/dE'$ and $d\Phi_G/dE'$ are the differential fluxes of cosmic (background) and gamma-ray (signal) emissions, and $R_B(E;E')$ and $R_G(E;E')$ are the telescope response functions to each of them (Sect. 4.2.1.1).

In practice, the response of the instrument R_B can be different for the background events coming from the *ON* and from the *OFF* regions, due to its dependence on the direction of the incoming particles within the observed FoV. Such discrepancies are measurable by the telescopes with relatively high precision, and the residual statistical and systematic uncertainties can be taken into account in the likelihood function through inclusion of the relevant nuisance parameters [1]. Indeed, this is done in the analysis presented in Chap. 5; in the following sections, however, for the characterization of the full likelihood method, it is assumedthat R_B is known with perfect precision and is equal in Eqs. (4.5) and (4.6). Still, in Sect. 4.2.6, the impact its uncertainties may have on the results is evaluated.

Apart from the shape of the spectral distribution, the given signal model $M(\boldsymbol{\theta})$ also predicts the expected number of detected events for a given effective observation time t_{eff}:

$$N_{\mathrm{EST}} = t_{\mathrm{eff}} \int_{E_{\min}}^{E_{\max}} P(E;M(\boldsymbol{\theta}))dE, \tag{4.7}$$

included in the full likelihood function (Eq. (4.2)) through the Poisson term.

Lastly, for the comparison of the full likelihood with the conventional analysis, it should be noted that their parameters relate as:

$$b = \frac{t_{\text{eff}}}{\tau} \int_{E_{\min}}^{E_{\max}} P_{\text{OFF}}(E)dE \tag{4.8}$$

and

$$g(\boldsymbol{\theta}) = t_{\text{eff}} \int_{E_{\min}}^{E_{\max}} P_{\text{ON}}(E; M(\boldsymbol{\theta}))dE - b. \tag{4.9}$$

Primary difference between the conventional and full likelihood methods can be illustrated with Fig. 4.1: both likelihoods are based on comparisons of the collected data with the predictions from the signal and background models. However, while the conventional approach integrates the spectral information in a pre-optimized energy range (for more details, see Sect. 4.2.2), and compares the expected and measured *number of events*, the full likelihood compares the expected and measured *energy distributions*, thus completely profiting from the potential presence of spectral features.

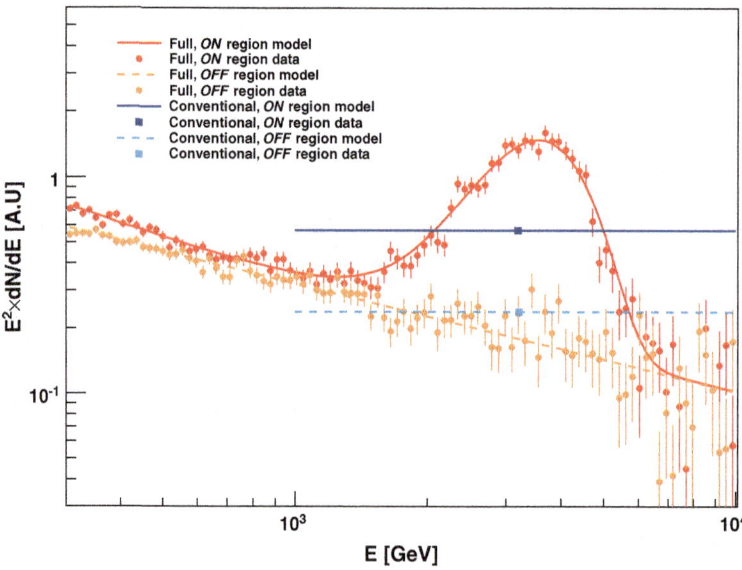

Fig. 4.1 Illustration of the advantage of the full likelihood method over the conventional one. *Red* and *orange lines* show the assumed spectral energy distributions of the *ON* and *OFF* regions, respectively, while the data points, with the same color code, represent the measured events (fine binning is used for the demonstration purposes only—the full likelihood is unbinned). The levels of *horizontal blue* and *cyan lines* correspond to the average value within the energy range considered in the conventional method, with points referring to the measurements. See the main text for more details

4.2 Characterization

In this section, in order to evaluate the performance of the proposed full likelihood concept, its properties are tested using fast simulations produced under a predefined set of conditions, and the results compared to those of the conventional method obtained under the exact same circumstances.

4.2.1 The Setup

4.2.1.1 Response Functions

As already described in Sect. 3.3.7.1, the response functions of an IACT for the background and gamma-ray events (R_B and R_G, respectively), depend on the properties of the instrument itself, on the effectiveness of the event reconstruction and on the analysis cuts applied in the given case.

The response function is usually presented as a combination of three components: the effective area $A_{\rm eff}(E', \hat{p}')$, angular $\Sigma(\hat{p}; E', \hat{p}')$ and energy $G(E; E', \hat{p}')$ reconstruction functions, where $\hat{p}'$ and $\hat{p}$ are the true and measured directions of the incoming particle. The spatial signatures may play a role for, e.g. galaxy clusters (Sect. 2.3.3, [2, 3]), as they can be predicted from the halo simulations (although, usually with great uncertainties). However, the analysis presented here is oriented towards source-candidates of angular size smaller or comparable to the typical PSF of the IACTs ($\sim$0.1°, Fig. 3.15)—hence, the contribution from the likelihood function dependent on the direction is not expected, and it can be integrated out. Therefore, the response function depends on the energy only:

$$R_{B,G}(E; E') = A_{\mathrm{eff}\,B,G}(E') \times G_{B,G}(E; E'). \tag{4.10}$$

For the characterization of the full likelihood method, as representative response function of the current-generation IACT, the approximation of the corresponding functions of the MAGIC Telescopes (Sect. 3.3.7.1) is used.

4.2.1.2 Spectral Functions

The spectral distributions needed for the full likelihood maximization are the ones of the background, $d\Phi_B/dE'$, and of the signal, $d\Phi_G/dE'$ (Eqs. (4.5) and (4.6)).

Background The background emission is produced by the cosmic rays, with a flux well described by a simple power law:

$$\frac{d\Phi_B}{dE'} = A_B E'^{-\alpha}, \tag{4.11}$$

with spectral index α and intensity A_B. In practice, however, the exact values of these parameters are not strictly necessary, since the needed value of $P_{\text{OFF}}(E)$ (Eq. (4.5)) is directly measured by the IACTs (or computed from MC simulations for projected instruments).

Signal For characterization purposes, two simple cases of signal emission are considered:

- a monochromatic line (L) at energy l and of intensity A_L:

$$\frac{d\Phi_G}{dE'} = A_L \delta(E' - l); \tag{4.12}$$

- power law (PL) of spectral index γ and intensity A_{PL}:

$$\frac{d\Phi_G}{dE'} = A_{PL} E'^{-\gamma}. \tag{4.13}$$

The convolution of the spectral and response functions yields the form of the PDF. As seen in Fig. 4.2, the original spectral shapes are modified by the imperfect instrument, with features like line being smoothed and hardness of the power law being altered.

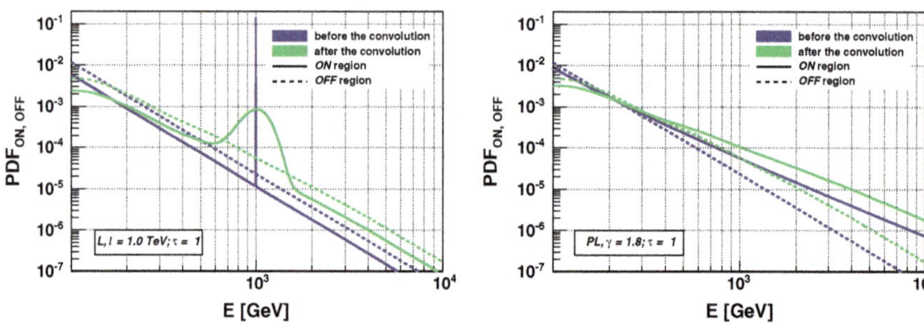

Fig. 4.2 Contributions of the *ON* (*full line*) and *OFF* regions (*dashed line*) to the PDF, before (*purple*) and after the convolution (*green*) of the spectral function with the response function of the telescope. *Left* a monochromatic line is smoothed and widened due to the finite energy resolution. *Right* the spectral slope of a power law-shaped signal is harder after the convolution. Shape of the background (*left* and *right*) is also affected by the response function

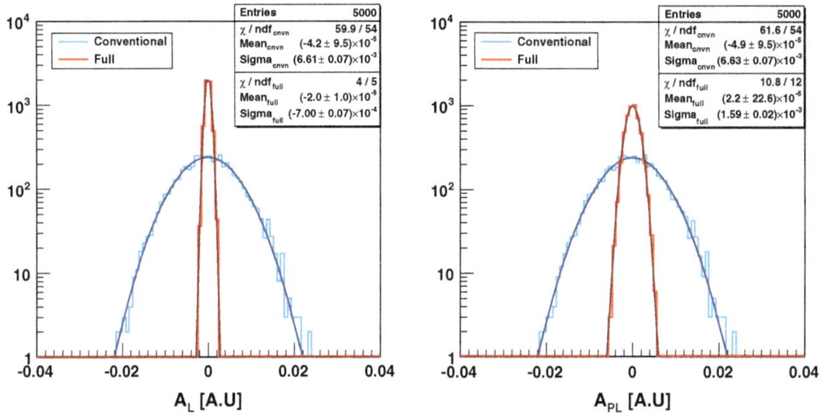

Fig. 4.3 Distribution of the free parameter values estimated by the conventional (*blue*) and full likelihood methods (*red*), for a line at energy l = 1 TeV (*left*) and power law of spectral slope γ = 1.8 (*right*) signal emission models. Test conditions are such that the expected parameter value is zero; results are obtained from 5000 fast-simulated experiments

4.2.1.3 Improvement Factor

In order to quantify the performance of the full likelihood with respect to that of the conventional approach, the *Improvement Factor* (*IF*) is defined as

$$IF(M(\boldsymbol{\theta})) = \langle CI_{\text{cnvn}}/CI_{\text{full}} \rangle, \tag{4.14}$$

i.e. the average ratio of the widths of the confidence intervals, CI_{cnvn} and CI_{full}, each calculated by the corresponding method. The CIs are estimated as two-sided, assuming a common c.l. and one unconstrained degree of freedom. In this work, the free parameter is chosen to be the signal intensity—for the characterization, that is A_L for the monochromatic line signal, and A_{PL} for the power law-shaped emission.

The concept of the Improvement Factor makes sense only if both the full likelihood and the conventional approach produce unbiased estimators. This extreme has been explicitly checked for several different models of signal emission and no indications for the presence of a bias have been found (Fig. 4.3).

By construction, the Improvement Factor is the *improvement in the sensitivity* of a given search expected from the use of full likelihood over the conventional approach. This is illustrated in Fig. 4.4: likelihood distributions, calculated by both methods, clearly indicate that, for the chosen confidence region, full likelihood is more sensitive as CI_{full} is more narrow than the CI_{cnvn}. For this particular case (PL signal with γ = 2.4), the Improvement Factor is $\sim$1.8[1]; in practice, this can be translated to, e.g. total

[1]The improvement by a factor of 1.8 is achieved without the range optimization for the conventional method (see Sect. 4.2.2).

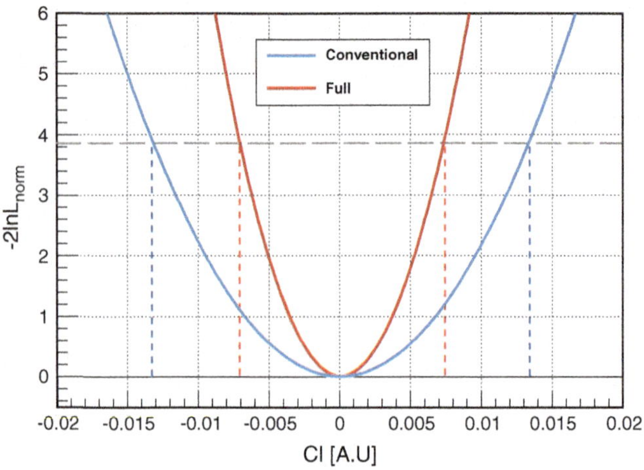

Fig. 4.4 Comparison of the *CI*s calculated by the conventional (*blue*) and the full likelihood (*red*) methods. The bounds of the *CI* (*vertical dashed lines*) correspond to a $-2lnL$ value above the minimum by a common quantity (3.84 in this case, corresponding to a 95 % c.l., *horizontal long-dashed line*). Narrower *CI* means better sensitivity. To make the comparison more obvious, the curve minima are shifted to the origin of the reference frame

observational time: the 80 % sensitivity gain achieved through the full likelihood method is equivalent to the results of the conventional analysis applied to about trice as much data.

Experimental conditions For the characterization of the method, unless specified otherwise, the confidence intervals are calculated with 95 % c.l., and their ratio averaged from 25 fast-simulated experiments. Each simulations consists of 10^5 events, randomly generated according to the PDF describing the expected background (i.e. the expected value of the signal intensity is zero). The background normalization is set to $\tau = 2$, and the considered energy range is between $E_{\mathrm{min}} = 100$ GeV and $E_{\mathrm{max}} = 10$ TeV. The maximization of the likelihood functions (Eqs. (4.1) and (4.2)) is performed using the `TMinuit` class incorporated in the framework of ROOT [4, 5].

4.2.2 Optimization of the Integration Range

In order to make a fair comparison of the performances of the full likelihood and conventional approaches, only the most constraining results from each method should be taken into the account. By definition, the full likelihood takes complete advantage of the signal spectral information; therefore, it makes sense to expect that maximal

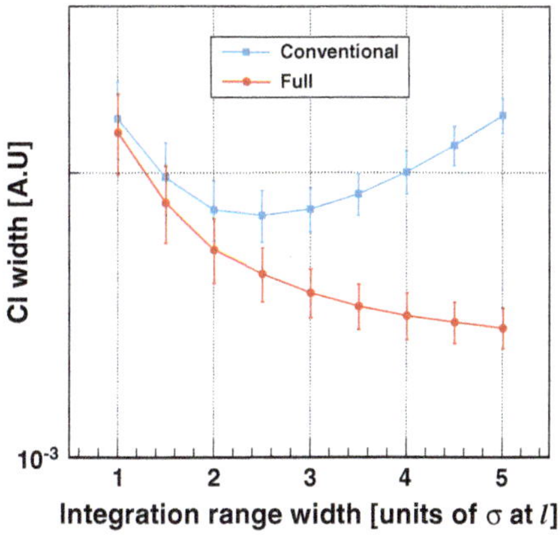

Fig. 4.5 Mean *CI*, calculated by the conventional (*blue*) and the full likelihood (*red*) methods, as a function of the integration range given in units of σ around the line energy $l = 1$ TeV. Error bars are the RMS of the *CI* distributions

sensitivity with this method is achieved when the whole energy range is considered. For the conventional concept, however, this does not have to be the case, especially if some distinctive features are expected in the spectra.

The following tests are devoted to the estimation of the performances of each of the two approaches for different energy integration ranges. For given model and method, the optimal integration range is the one resulting in the best sensitivity.

Line In the case of a monochromatic line signal model, the sensitivity of likelihood analyses is optimized by restricting to those events in the vicinity of the peak. Figure 4.5 shows the *CI* widths of the conventional and full likelihood approaches, as a function of the integration range width centered at l. Given how the ability of the instrument to distinguish characteristic spectral features, like lines, is determined by its resolution at the corresponding energies, the integration range width is expressed in units of σ.

As expected, the full likelihood is best favored when the entire energy range is considered (out of the range in Fig. 4.5). On the other hand, the conventional approach is most sensitive for a particular, limited range: in the case of a line at $l = 1$ TeV, the optimal integration range is 2.5σ wide.

Power Law For the power law-shaped signal models, the optimization is done by fixing one integration limit while varying the other. Figure 4.6 shows the mean *CI* for each method, for a signal model of spectral slope $\gamma = 1.8$ and an integration range of fixed E_{min} or fixed E_{max}. Again, in both cases, the full likelihood is best favored when the entire energy range is considered. As for the conventional approach, the scenario with fixed E_{max} and optimized E_{min} yields the best sensitivity.

The Improvement Factor values given in the following sections are always calculated from the most constraining upper limits of both methods, using the entire energy range for the full likelihood and the optimized one for the conventional approach.

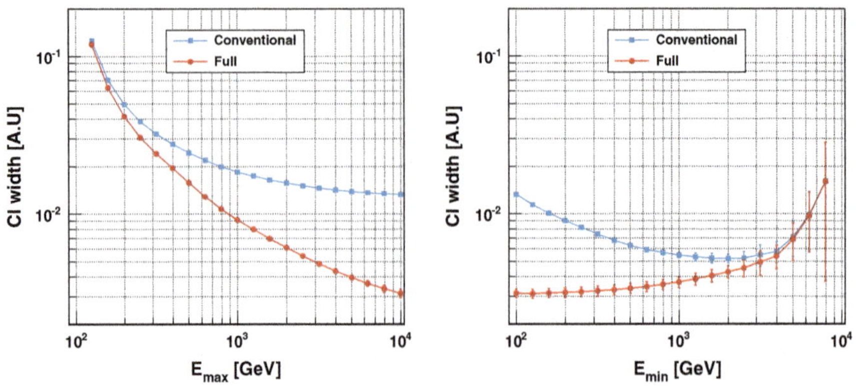

Fig. 4.6 Mean *CI*s, calculated by the conventional (*blue*) and full likelihood approach (*red*), as a function of integration range when $E_{\min}$ (*left*) or $E_{\max}$ (*right*) is fixed. The considered signal model is a *PL* of spectral slope $\gamma = 1.8$. Error bars are the RMS of the *CI* distributions

4.2.3 Improvement Factor for Different Signal Models

The following tests compare the sensitivities of the conventional and full likelihood methods for various power law and line-shaped signal models.

Line For the line models the Improvement Factor values range between 40 % and 65 %, depending on the line energy l (Fig. 4.7). The dashed line refers to the optimal integration range width for the conventional approach (in units of σ at l). It is interesting to note that this width is almost constant for all the models and of order of $2.5-3\sigma$.

Power Law The Improvement Factor values for *PL* signal models depend on the slope of the spectral index γ (Fig. 4.8). The softer the spectrum, the lower the gain the full likelihood provides over the conventional approach. For instance, for the chosen characterization setup, for the case when $\gamma \approx 3.6$, the shapes of signal and background differential rates are very alike, and therefore the improvement one achieves from the use of the full likelihood is almost negligible. For harder spectral slopes the advantage of the full likelihood over the conventional one is far more significant, with Improvement Factor values of up to ∼70 %. The dashed line indicates the optimized value of $E_{\min}$ for the conventional method: for expected signal emissions of harder spectral indices, that dominate over the background radiation at higher energies, the conventional approach is optimized for the upper end of the energy range. For increased γ, differences between the signal and background concentrate at lower energies, so integration of the entire energy range is preferred.

Additional Features The spectral shape of the signal can be further elaborated by including additional features of physical interest. For instance, there can be a sharp cutoff in the spectral distribution, smoothed by the response function of the detector. Figure 4.9 considers the case of *PL* models with different spectral slopes γ that all have a cutoff at a fixed energy of 1 TeV. In the presence of a cutoff, the Improvement

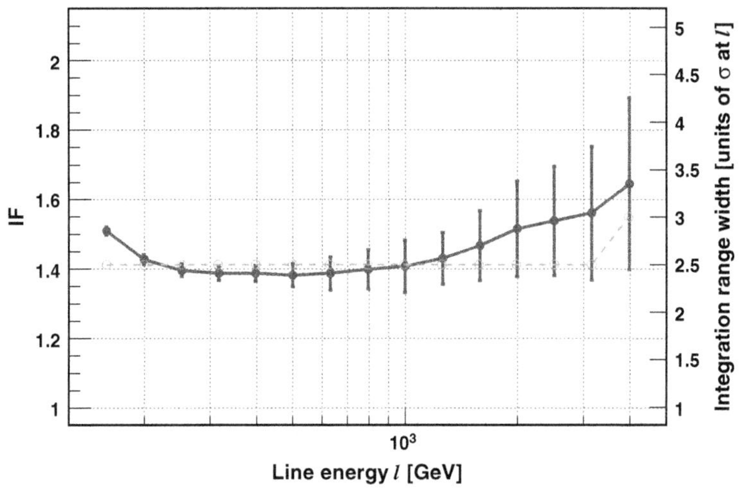

Fig. 4.7 Improvement Factor for different L signal models (*full line*). Also shown are the optimal integration range widths for the conventional approach for the considered models (*dashed line, right-hand axis*). Error bars are the RMS of the IF distributions

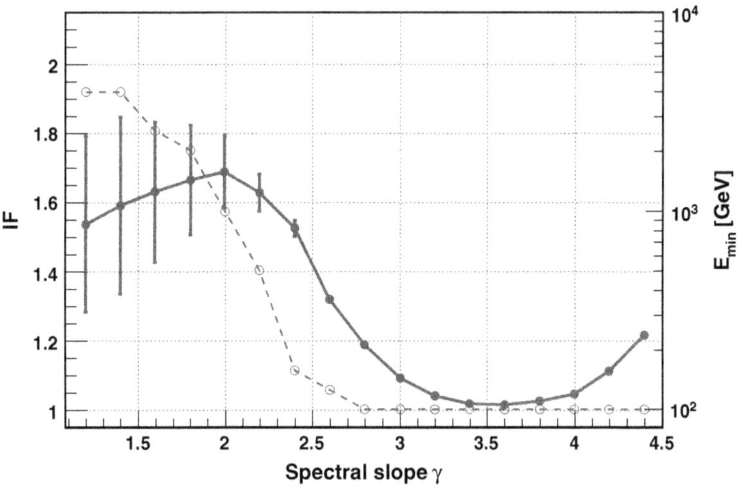

Fig. 4.8 Improvement Factor for different PL signal models (*full line*). Also shown are the optimal values of $E_{\rm min}$ for the conventional approach for the considered models (*dashed line, right-hand axis*). Error bars are the RMS of the IF distributions

Factor is lower than in the case of uninterrupted PL emission. This is especially noticeable for those signal models that dominate at high energies ($\gamma > \alpha$), since their distinction from the background is partially erased by the cutoff. For the softer spectra, this effect is not that evident, as for those cases signal is more distinguishable from the background at lower energies, i.e. well below the cutoff.

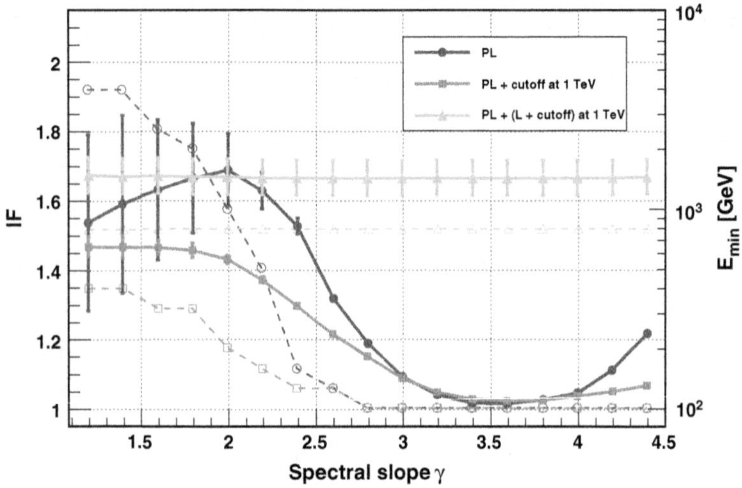

Fig. 4.9 Improvement Factor as a function of spectral slope γ for different signal models (*full line*). Also shown are the optimal values of $E_{\min}$ for the conventional approach for the considered models (*dashed line, right-hand axis*). Error bars are the RMS of the IF distributions

The dependency of the Improvement Factor value on the cutoff energy is shown on Fig. 4.10: for hard spectra, the higher the cutoff, the greater the gain from the use of the full likelihood. Nevertheless, the loss of signal events due to the cutoff always keeps the Improvement Factor below the value of that for the corresponding *PL* model without the cutoff. As for the softer spectra, the improvement is enhanced

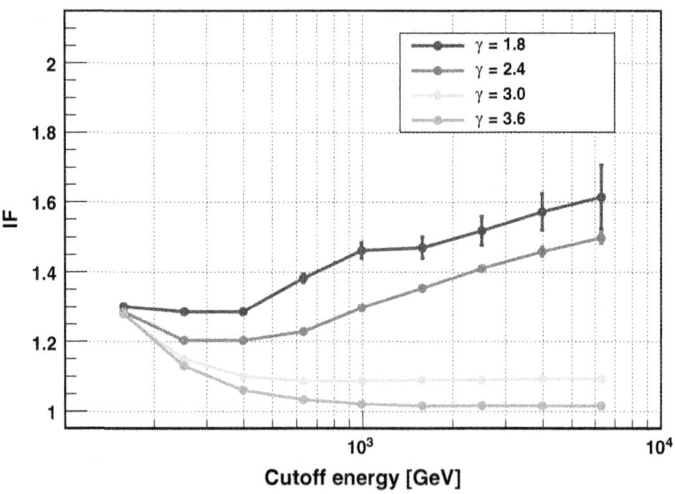

Fig. 4.10 Improvement Factor as a function of cutoff energy for different *PL* signal models. Error bars are the RMS of the IF distributions

by low-energy cutoffs to levels comparable to those obtained for spectral lines at similar energies.

Lastly, the effect of adding a line to a power law-with-the-cutoff spectral distribution is examined. For such models, the overall signal intensity is taken as the free parameter, while the individual amplitudes of the power law (A_{PL}) and line contributions (A_L) are set in such a manner that the integrated areas corresponding to those emissions in the PDF are equal. As shown in Fig. 4.9, the presence of a line at the same energy as the cutoff (l = 1 TeV) significantly boosts the Improvement Factor value, especially for soft spectra. Its contribution is obvious from the optimal E_{min} distribution as well: regardless of the value of γ, the most constraining limits from the conventional method are achieved when E_{min} is just below the line, seeing how this feature is the one dominating the Improvement Factor value.

4.2.4 Stability

Previous sections dealt with variations of the signal model; this one examines the dependence of the Improvement Factor on the experimental parameters. Table 4.1 summarizes the results obtained assuming different values of parameters not affiliated with the model itself, but, rather with the observational setup, characteristics of the instrument and choice of the analysis cuts. Also considered is the effect a presence of a signal in the data may have on the Improvement Factor values.

For the majority of the considered settings, the observed improvement variations are of order 1–2 %. Exceptions are described in more detail below.

Background Normalization in first approximation, can simply be the number of chosen *OFF* regions (in the observational scheme or in the analysis (Sects. 3.2.1, and 3.3.6)) for the estimation of the residual background in the *ON* region. More *OFF* regions means larger statistics, and consequently, more constraining results from the

Table 4.1 Dependence of the Improvement Factor on different experimental parameters for three different representative signal models

Parameter	Variation range (units of the parameter)	IF		
		PL, $\gamma = 1.8$	*PL*, $\gamma = 3.6$	*L*, $l = 1$ TeV
τ	1–5	1.91–1.47	1.02–1.01	1.63–1.26
Number of events	5×10^4 – 5×10^6	1.66–1.62	1.03–1.02	1.43–1.41
σ [% of σ_{MAGIC}]	50–500	1.65–1.66	1.01–1.11	1.37–3.23
E_{max} [TeV]	10–50	1.65–1.82	1.01–1.02	1.40–1.41
S [standard deviations]	0–5	1.65–1.75	1.01–1.01	1.40–1.42

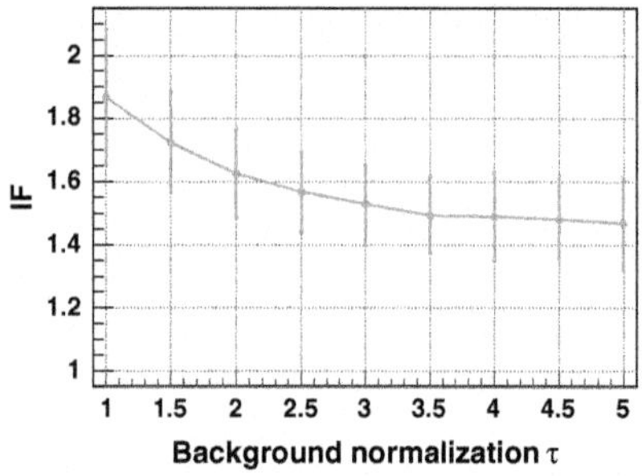

Fig. 4.11 Improvement Factor dependence on the background normalization τ, for a *PL* signal models of spectral slope $\gamma = 1.8$. Error bars are the RMS of the IF distributions

analyses. Figure 4.11 shows the dependence of the Improvement Factor on the value of τ, for the power law-shaped spectra of $\gamma = 1.8$. As it can be seen, for hard spectra, the greater the τ, the lower the gain provided by the full likelihood method. The same conclusion applies to the line signal models (Table 4.1). On the other hand, soft power law-shaped signals are not significantly affected by the τ.

Number of Events chosen for the characterization (10^5 events) translates to ~200 h of data, assuming the described setup (Sect. 4.2.1). This number, however, strongly depends on the chosen instrument and applied analysis cuts (in particular, on the energy threshold of the analysis (Sect. 3.3.5.3)). Nevertheless, the number of the events included in the likelihood functions does not play a significant part in the overall Improvement Factor value (Table 4.1).

Energy Resolution of a given instrument reflects its ability to distinguish characteristic spectral features. It is therefore expected that the value of the Improvement Factor depends on the σ of the detector. For the line signal models, as seen from Fig. 4.12, the higher the σ, the greater the improvement. It must be clarified, however, that this does not imply that a poor resolution yields more constraining results (for neither of the studied likelihoods), but that the advantage of the full likelihood over the conventional one is more significant. Regarding the power law-shaped emissions, as they have no sharp features, the Improvement Factor values do not depend significantly on σ (Table 4.1).

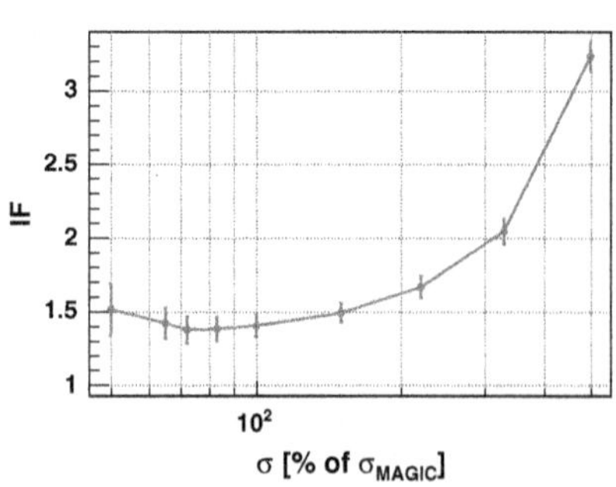

Fig. 4.12 Improvement Factor dependence on the energy resolution σ of the instrument, for *L* signal model centered at energy l = 1 TeV. Error bars are the RMS of the IF distributions

Energy Range of the given analysis, parameterized by E_{max}, depends on the telescope and the chosen cuts. For the line signal models, the expansion of the total energy range only adds more background, and the Improvement Factor value is not significantly affected (Table 4.1). The same goes for soft power law-shaped spectra, that are distinguished from the background only at lower end of the energy range. Hard spectral emission, however, dominates at high energies, and the increase of the energy range also means a greater Improvement Factor.

Signal Intensity can also influence the Improvement Factor value. The presence of a signal sufficiently strong to be detected (using the full likelihood method) with S of up to 5 standard deviations, yields for hard power law models a sensitivity improvement of up to 10 %. This can be understood from the following: for the conventional method, the optimization of the integration range was performed assuming that the sample contains no signal. Thus, as a consequence, if there is an excess in the data, some gamma-ray events may be lost. For the full likelihood, on the other hand, everything is included in the calculations by default (Sect. 4.2.2). As for the line and soft power law models, signal presence causes no relevant change in the Improvement Factor value (Table 4.1).

4.2.5 Robustness

The robustness of the full likelihood is evaluated by assuming that the response function of the instrument is not precisely known. For the following tests, the events are simulated with one response function, R_0, while a different (wrong) one, R_W is used for the likelihood maximization. Data are generated so that they contain a gamma-ray signal of intensity that yields a 5σ detection for $R_0 = R_W$ with the full likelihood. The studies below describe how the significance of the detection degrades when $R_0 \neq R_W$.

Effective Area Let the A_{eff} function, assumed to describe the response of the detector, be shifted by a fixed energy with respect to the actual A_{eff}. As a result, the sensitivity of the detection is decreased, for power law-shaped signal models, by up to 5 % for a 50 GeV shift. The effect on the line-like models is not significant (less than 1 %, Fig. 4.13).

Energy Resolution Next, the case of unknown σ is considered: a factor 2 mistake in the estimate of the energy resolution leads to up to $\sim$10 % worst sensitivity for line-like models (Fig. 4.13). For power law spectra, a σ wrong by the same factor has no significant effect—less than 1 % sensitivity decline.

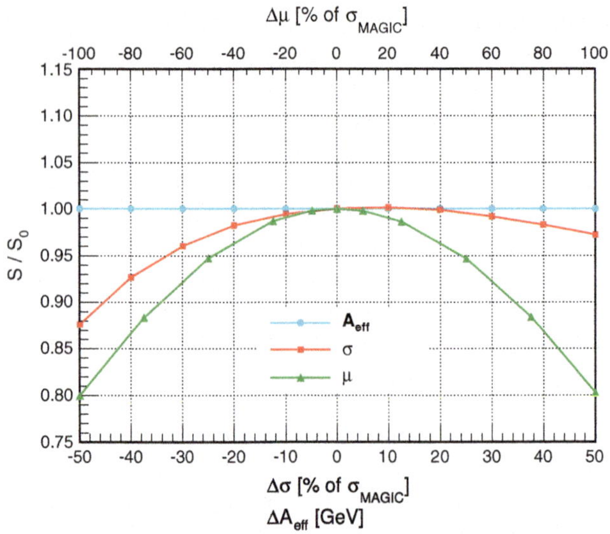

Fig. 4.13 Relative decrease of sensitivity of the full likelihood method as a result of a not-precisely-known effective area (*blue*), energy resolution (*red*) or energy bias (*green*). The x-axes mark the uncertainties: ΔA_{eff} (*bottom*) is the absolute shift in energy of the A_{eff} function. Same scale is shows the relative change of σ with respect to the σ_{MAGIC}. Shift of μ value is given in % of σ_{MAGIC} (*top*). The considered signal model is a line with l = 1 TeV

Energy Bias Lastly, different μ functions are presumed for the simulation and for the likelihood analysis. Findings show that, for μ values shifted from the actual ones by 1σ at a given energy, the sensitivity of the analysis decreases ~5 % for line-like models (Fig. 4.13). If the shift is 2σ, the decline is up to 20 %. This means that, for instance, when searching for a line in the spectrum, steps wide as up to 1σ can be made in the scan without risking a significant sensitivity degradation. For the power law-shaped spectra no significant sensitivity loss has been found.

Having in mind that even under these extreme and conservative conditions, the worsening in the sensitivity of the full likelihood method is still smaller than the gain its use provides, it may be concluded that this method is robust.

4.2.6 Background

As mentioned in Sect. 4.1.2, the residual background in *ON* region may be estimated within some uncertainties. This section evaluates the effect that these uncertainties can have on the performance of the conventional and full likelihood methods.

First, the energy-dependent differences between the R_B functions for *ON* (Eq. (4.6)) and *OFF* regions (Eq. (4.5)) are considered. For the parametrization of such discrepancies, an extra power law term multiplying the first integral in Eq. (4.6)

is added. Its index is introduced in the likelihood functions as a nuisance parameter, with Gaussian probability distribution of mean 0 and width 0.04 (so that a maximum deviation of 5 % at any energy is reached). This results in a sensitivity decrease for both the full likelihood and the conventional methods, but more drastically for the latter one: for the case of line-like models, as well as for hard power law-shaped spectra, results from the conventional approach are up to ∼50 % less constraining. For the full likelihood, the corresponding sensitivity losses are smaller: ∼5 % for line and ∼25 % for power law signal models. Soft power law-shaped spectra are not affected (less than 1 %), for either of the analysis methods.

The case of global (normalization) differences between R_B functions considered in Eqs. (4.5) and (4.6) is examined by treating τ as a nuisance parameter, with a Gaussian probability distribution of 5 % width. This leads to significant sensitivity losses for the conventional method: ∼30 % for the line-like and ∼10 % for the hard power law signals. The full likelihood is again far more robust, exhibiting almost negligible worsening—less than 2 % for both kinds of signal models. On the other hand, soft power law models result problematic for both approaches, especially when the spectral shape of the signal is similar to that of the convoluted background. The conventional approach suffers from an up to factor ∼8 worse sensitivity, also for all softer models. For the full likelihood this is less pronounced (up to a factor ∼4 sensitivity degradation), and the power of the method is recovered as soon as the shape of the expected signal distribution becomes different from that of the background. This is caused, for both approaches, by high correlation (up to 0.99) between τ and signal intensity, when signal and background are of similar spectral shapes. For other signal models the correlation is low, due to the energy range optimization applied in the conventional approach (Sect. 4.2.2) and the presence of the spectral term in the full likelihood expression (Eq. (4.2)).

4.3 Sensitivity of the Full Likelihood Method for Dark Matter Searches

Until now, the performance of the full likelihood has been characterized in a rather general way, by assuming generic spectral shapes. In this section, its sensitivity is explored for specific dark matter models, for the observations with MAGIC and CTA. The sensitivity is expressed as the achieved upper limit to the value of $\langle\sigma_{\mathrm{ann}} v\rangle$, which is taken to be the free parameter in the maximization of the likelihood.

Bringmann, Doro and Fornasa (2009, [6]) made observability predictions (requiring a 5σ detection in 50 h) for two dSph galaxies, Draco and Willman 1, for the case of several mSUGRA (Sect. 1.3.1.1) benchmark (BM) models [7], for MAGIC stereo and CTA. However, as neither of those instruments was operational at the time, the response functions attributed to each system were rather simplified and slightly optimistic. For the calculations, they relied on the conventional likelihood approach, and made two studies: one, for which E_{min} is the actual energy threshold of the analysis (70 GeV for MAGIC, 30 GeV for CTA), and the other, for which E_{min} is optimized

Table 4.2 Characteristics of the studied BM models (mass m_χ and predicted annihilation cross section today $\langle\sigma v\rangle|_{v=0}$), together with the $\langle\sigma v\rangle$ upper limits calculated with full likelihood method ($\langle\sigma v\rangle_{\text{full}}$), for Willman 1 observations with MAGIC and CTA

	BM	I'	J'	K'	*BM3*	*BM4*
	m_χ (GeV)	141	316	565	233	1926
	$\langle\sigma v\rangle\|_{v=0}$ (cm^3 s^{-1})	3.6×10^{-27}	3.2×10^{-28}	2.6×10^{-26}	9.2×10^{-29}	2.6×10^{-27}
MAGIC	$\langle\sigma v\rangle_{\text{full}}$(cm^3 s^{-1})	5.65×10^{-23}	1.01×10^{-23}	3.91×10^{-23}	7.21×10^{-25}	2.87×10^{-23}
	IF_1	1.62	3.64	1.23	4.14	2.10
	IF_2	1.57	1.80	1.23	1.89	2.10
	$\langle\sigma v\rangle_1^{2009}$ (cm^3 s^{-1})	3.31×10^{-23}	2.21×10^{-23}	2.57×10^{-23}	1.38×10^{-24}	4.94×10^{-23}
	$\langle\sigma v\rangle_2^{2009}$ (cm^3 s^{-1})	2.23×10^{-23}	5.12×10^{-24}	2.57×10^{-23}	7.45×10^{-25}	9.62×10^{-24}
CTA	$\langle\sigma v\rangle_{\text{full}}$ (cm^3 s^{-1})	1.39×10^{-23}	1.91×10^{-24}	8.39×10^{-24}	1.35×10^{-25}	4.82×10^{-24}
	IF_1	1.48	5.18	1.58	6.62	3.81
	IF_2	1.48	1.65	1.58	1.61	3.81
	$\langle\sigma v\rangle_1^{2009}$ (cm^3 s^{-1})	5.46×10^{-24}	4.48×10^{-24}	5.20×10^{-24}	5.98×10^{-25}	4.42×10^{-23}
	$\langle\sigma v\rangle_2^{2009}$ (cm^3 s^{-1})	3.60×10^{-24}	7.68×10^{-25}	3.38×10^{-24}	1.29×10^{-25}	1.59×10^{-24}

Also quoted are the Improvement Factors obtainable from the full likelihood with respect to the conventional approach, computed according to the prescription presented in [6]: IF_1 is calculated for an integration range (for the conventional method) from energy threshold to m_χ, while for IF_2 the integration is done from optimized lower limit to m_χ. For comparison, the corresponding $\langle\sigma v\rangle$ bounds from [6] are quoted as well, for the considered integration ranges ($\langle\sigma v\rangle_{1,2}^{2009}$)

for based on the spectral shapes of the considered models and the sensitivity curves of the instruments. In both cases, $E_{\max}$ is selected as the dark matter particle mass m_χ.

To quantify the improvement the full likelihood yields over the conventional one, the sensitivities for both methods are computed following the prescription laid out in [6]: considering the same dark matter candidate source (Willman 1), the same BM models as signal emission, and the same observatories (but with more realistic response functions: the actual one of MAGIC (Sect. 3.3.7.1) and one of the latest simulations of the CTA response [8]). The results are summarized in Table 4.2, together with the basic characteristics of each of the studied BM models and the corresponding $\langle\sigma_{\rm ann}v\rangle$ values predicted in [6]. The Improvement Factors IF_1 and IF_2 represent the gain the full likelihood provides over the conventional approach (both calculated in this work), for the two cases of integration ranges considered in [6].

The lowest Improvement Factors (although of values higher than 25 %), are obtained, for both MAGIC and CTA, for the practically featureless, soft spectra of the model K', as well as for the model I' that has a cutoff at low energies (Fig. 4.14). On the other hand, the greatest improvements are achieved in the case of the model $BM4$, characterized by the a m_χ and hard spectrum. Models with particularly significant internal bremsstrahlung contributions, J' and $BM3$, also show significant gain from the use of the full likelihood—more than 60 %. Despite these high improvements, however, the estimated $\langle\sigma_{\rm ann}v\rangle$ limits are still ~4 and ~3 orders of magnitude away, for MAGIC and CTA, respectively, from constraining these models.

The fact that results for $\langle\sigma_{\rm ann}v\rangle$ calculated with the full likelihood are a factor ~2 less constraining from the conventional limits presented in [6] ($\langle\sigma v\rangle_{1,2}^{2009}$, Table 4.2), can be understood by taking into the account that the latter were obtained assuming a somewhat idealized situation, with perfectly known background and with flat, optimistic response functions, while this work considers circumstances of the real experiment and the actual (or latest from the simulations) response of the detectors.

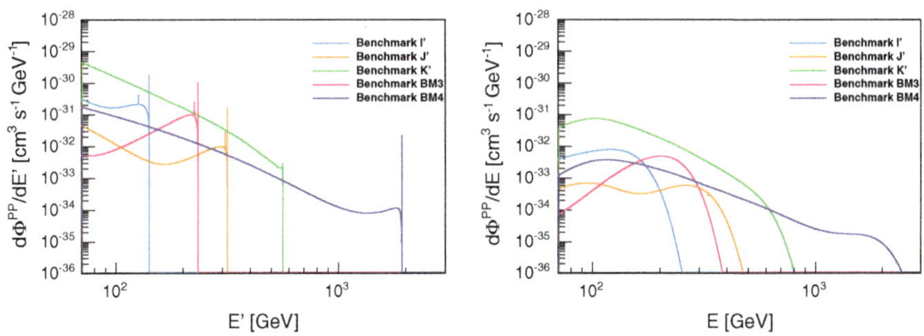

Fig. 4.14 The considered BM models before (*left*, adapted from [6]) and after the convolution with the MAGIC response function (*right*)

4.4 Overview of the Full Likelihood Method

In this chapter, the concept of the full likelihood method has been introduced, and its performance characterized. As shown, this method is constructed to take the maximal advantage of the spectral information, and almost solely through the inclusion of the a priori knowledge on the expected gamma-ray spectrum in the likelihood, it achieves significantly better sensitivity than the conventional method.

Furthermore, it has been demonstrated that the full likelihood is unbiased and robust; the sensitivity gain from its use is rather independent on other analysis characteristics, like the background estimation or signal-to-background discrimination. As a result, the full likelihood method can be combined to any other analysis development aimed at further sensitivity enhancements.

The focus of this work is set on the indirect searches for dark matter annihilation and decay signals with IACTs. This is reflected in the specific form of the likelihood function (Eq. (4.2)), determined by the fact that IACT observations are pointed, cover a relatively narrow FoV, and are dominated by background events. Although, the knowledge of the Author, never before used for IACTs, this concept is a well known analysis method, successfully applied in other fields, including dark matter searches with different techniques and instruments. For instance, a similar approach is employed in the direct detection experiments, like XENON100 [9], and even more extensively, in the indirect searches for dark mater signals in gamma rays by the Fermi-LAT (see, e.g., [10]).

The proposed method is sufficiently general to be used in studies of other IACT physics cases, the only condition being that a prediction about the expected spectral distribution can be made. For instance, it can be employed in the search of the active galactic nuclei spectra for signatures induced by the oscillations of gamma-rays into axion-like particles in the presence of intergalactic magnetic fields [11]. This case, however, would require the a priori assumptions on the active galactic nuclei emission and effects of gamma rays interacting with the extragalactic background light.

Another very important characteristics of the full likelihood (and any likelihood-based analysis) is that it allows a rather straightforward combination of the results obtained by different instruments and from different targets. For a given model ($M(\boldsymbol{\theta})$) and N_{inst} different instruments (or measurements), a global likelihood function can be simply written as:

$$\mathscr{L}(M(\boldsymbol{\theta})) = \prod_{i=1}^{N_{\text{inst}}} \mathscr{L}_i(M(\boldsymbol{\theta})). \tag{4.15}$$

This approach eliminates the complexity required for a common treatment of data and response functions of different telescopes or analyses: within the likelihood scheme, the details of each experiment do not need to be combined or averaged. The only necessary information is the value of the likelihood, expressed as a function of the free parameter of a given model for different instruments.

Considering all of the above, the full likelihood method is alogical choice for the analysis in the dark matter searches with IACTs: following chapters present the results of this approach applied to the actual MAGIC observations, as well as the predictions on the sensitivity achievable with the future CTA.

References

1. W.A. Rolke, A.M. López, J. Conrad, Limits and confidence intervals in the presence of nuisance parameters. Nucl. Instrum. Meth. A **551**, 493–503 (2005). arXiv:physics/0403059
2. A. Pinzke, C. Pfrommer, L. Bergström, Prospects of detecting gamma-ray emission from galaxy clusters: cosmic rays and dark matter annihilations. Phys. Rev. D **84**, 123509 (2011). arXiv:1105.3240
3. M.A. Sanchez-Conde, Dark matter searches with Cherenkov telescopes: nearby dwarf galaxies or local galaxy clusters? JCAP **12**, 011 (2011). arXiv:104.3530
4. F. James, MINUIT. Function minimization and error analysis, reference manual version 94.1, CERN Program Library Long Writeup D506, CERN, Geneva, Switzerland (1994)
5. The ROOT Team, TMinuit. http://root.cern.ch/root/html/TMinuit.html (January, 2013)
6. T. Bringmann, M. Doro, M. Fornasa, Dark matter signals from Draco and Willman 1: prospects for MAGIC-II and CTA. JCAP **01**, 016 (2009). arXiv:0809.2269
7. M. Battaglia et al., Updated post - WMAP benchmarks for supersymmetry. Eur. Phys. J. C **33**, 273–296 (2004). arXiv:hep-ph/0306219
8. K. Bernlöhr et al., Monte Carlo design studies for the Cherenkov Telescope Array. Astropart. Phys. **43**, 171–188 (2013). arXiv:1210.3503
9. The XENON DarkMatter Project, XENON100 Experiment. http://xenon.astro.columbia.edu/XENON100_Experiment (April, 2013)
10. M. Ackermann et al., Constraints on dark matter annihilation in clusters of galaxies with the Fermi Large Area Telescope. JCAP **05**, 025 (2010). arXiv:1002.2239
11. M.A. Sanchez-Conde et al., Hints of the existence of axion-like particles from the gamma-ray spectra of cosmological sources. Phys. Rev. D **79**, 123511 (2009). arXiv:0905.3270

Chapter 5
Dark Matter Searches in Dwarf Spheroidal Galaxy Segue 1 with MAGIC

This chapter presents the results of this work—the search for signals of dark matter in dSph galaxy Segue 1. The long-term observational campaign was carried out with the MAGIC Telescopes between January 2011 and February 2013. With 157.9 h of good-quality data, this is the deepest survey of any dwarf galaxy by any IACT so far. No significant gamma-ray excess has been found. Using the full likelihood approach, limits have been set assuming different models of dark matter annihilation and decay. At the moment, those are the most constraining limits from dSphs observations with IACTs.

This chapter begins with the motivation behind the choice of Segue 1 as a suitable dark matter target, followed by the details of its observations with MAGIC and the subsequent analysis. Lastly, results from the full likelihood method and their interpretation in the light of various dark matter models are presented.

5.1 Segue 1 as Target for Dark Matter Searches

5.1.1 Introduction to Dwarf Spheroidal Galaxies

In the context of ΛCDM, dark matter structures form by hierarchical collapse of small overdensities and are presumed to extend in mass down to the scale of the Earth, or even below [2]. Some of these substructures may have attracted enough baryonic matter along their evolutionary road to commence the star creation. This hypothesis has been used to explain the dSph satellite galaxies of the Milky Way, proposed to have formed within the substructures of the Galactic dark matter halo [3, 4].

The first galaxies identified as dSphs were Sculptor and Fornax, in 1938 [5]. By the end of the 20th century, due to their faint nature, only seven more of these objects were discovered. Given how the number of known dSphs was orders of magnitude lower that the one predicted by the N-body simulations (the so-called *missing satellite*

J. Aleksić, *Optimized Dark Matter Searches in Deep Observations of Segue 1 with MAGIC*, Springer Theses, DOI 10.1007/978-3-319-23123-5_5

problem, [6]), it was reasonable to assume that the existing census of Milky Way satellites was incomplete, and that new detections may be expected from the more sophisticated photometric surveys.

Indeed, since 2004, with the advent of the Sloan Digital Sky Survey (SDSS, [7]), a dozen of new dSph galaxies have been identified. The newly-discovered dSphs, however, are not just significantly fainter than their predecessors, which was to expect, but also posses certain properties more akin to the globular clusters than to the dwarfs. It is why the distinction is made between the *classical* dSphs, detected in the pre-SDSS era, and the *ultra-faint* ones, discovered since 2004.

The classical dSphs are by now well-established sources with hundreds of identified member stars, located at distances from 70 to 250 kpc (Table 5.1). They are more extended and more luminous than their ultra-faint counterparts, with half-light radii typically of the order of few hundreds of parsecs, and luminosities spread over nearly two orders of magnitude (10^5–10^7 $L_\odot$). Thanks to their many bright giant constituents, velocities of their member stars can be measured to a precision of a few km/s, or less [8].

On the other hand, all of the known ultra-faint dSphs have total luminosities lower than the faintest classical satellite ($L \leq 10^5$ $L_\odot$). Their stellar population is

Table 5.1 Compilation of properties of Milky Way satellites: distance d, absolute magnitude M_V, luminosity L, half-light radius r_h, velocity dispersion $\langle\sigma^2\rangle$ and the year of discovery

Name	d [kpc]	M_V	L [$10^6 L_\odot$]	r_h [pc]	$\langle\sigma^2\rangle$ [km/s]	Discovered
Sculptor	80	−9.8	2.15	283	9.0	1938
Fornax	138	−13.1	15.5	710	10.7	1938
Leo I	250	−11.9	4.79	251	9.0	1950
Leo II	205	−10.1	0.58	176	6.6	1950
Draco	80	−9.4	0.26	221	10.1	1954
Ursa minor	69	−8.9	0.29	181	11.5	1954
Carina	101	−9.4	0.43	250	6.4	1977
Sextans	86	−9.5	0.50	695	7.1	1990
Willman 1	38	−2.7	1.0×10^{-3}	25	4.0	2005
Ursa major I	106	−5.5	1.4×10^{-2}	319	7.6	2005
Ursa major II	32	−4.2	4.0×10^{-3}	149	6.7	2006
Hercules	138	−6.6	3.6×10^{-2}	330	5.1	2006
Leo IV	158	−5.0	8.7×10^{-3}	206	3.3	2006
Canes Venatici I	224	−8.6	2.3×10^{-1}	564	7.6	2006
Canes Venatici II	151	−4.9	7.9×10^{-3}	74	4.6	2006
Coma Berenices	44	−4.1	3.7×10^{-3}	77	4.6	2006
Segue 1	23	−1.5	3.4×10^{-4}	29	4.3	2006
Bootes I	60	−6.3	8.4×10^{-2}	242	9.0	2004

Classical dSphs are above the horizontal line, while the ultra-faints are below. The table only shows galaxies with well-measured kinematic data, for which the mass can be determined from $\langle\sigma^2\rangle$ and r_h. Compiled from [9, 10]

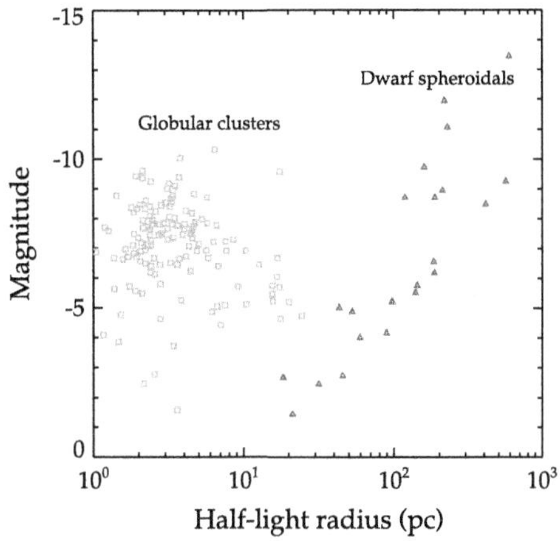

Fig. 5.1 Relationship between the absolute magnitude and the half-light radius for Milky Way globular clusters (*red squares*, left side) and dSphs (*blue triangles*, right side). The ultra-faint satellites appear as the *blue triangles* in the lower portion of the figure. Taken from [1]

small, consisting of only tens of very faint stars (with typical apparent magnitudes of 20−21). This affects the precisions of velocity dispersion measurements (uncertainties are 2−3 km/s, which is about a factor of two of the intrinsic velocity dispersion of the system), as well as the total luminosity estimation (which can be inflated if not all of the interloping stars are removed from the dSph sample [11]). The half-light radii of some ultra-faints are in the 30−100 pc range (Table 5.1), which is more typical for globular clusters than for galaxies (Fig. 5.1). Furthermore, velocity dispersions of both dwarfs and clusters are similar (5−15 km/s, [12]), and there is an overlap in their absolute magnitudes. With ultra-faints behaving as 'in-between' the classical dSphs and globular clusters, it is very important to correctly derive the properties and make classification of the newly discovered object as either a globular cluster, for which dynamical mass within the half-light radius is dominated by stars, or as a dSph galaxy, for which dark matter will be the principal component. Distinction between the two kinds of systems is based on their different chemical compositions: the globular clusters are on average metal-rich ($[Fe/H] > -2$) and show a small amount of internal metallically spread, unlike the metal-poor dSphs [13]. In addition, their evolutionary histories differ: the clusters largely formed their star population in a single burst, while in a galaxy the formation is usually extended, with member stars belonging to different stages of dSph evolution [14].

Some of the known classical and ultra-faint dSph galaxies with well measured kinematic data, together with their basic properties, are listed in Table 5.1.

5.1.2 Dwarf Spheroidal Galaxies as the Dark Matter Candidates

From measurements of velocities of member stars in dSphs, it has been estimated that their M/L ratio is of order of hundred, or even thousand $M_\odot/L_\odot$ for some ultra-faint dwarfs, which makes these galaxies the most dark matter dominated objects (known) in the Universe [15, 16]. Furthermore, given their extreme total M/L ratios, it is unlikely that the luminous material has altered the distribution of dark matter in these systems [17]. The gamma-ray background expected from dwarfs is very low: the poor baryonic content disfavors presence of conventional gamma-ray emitters (such are supernova remnants, pulsars or binary systems); there are no indications of recent star formation history; there is little or no gas in their interstellar medium to serve as target material for cosmic rays [18]. Locations of dSphs are well known and many of these galaxies are within 100 kpc distance from Earth. Finally, some of the dwarfs are located at high galactic latitudes, where contamination from the Galactic background is subdominant.

Given some of their above mentioned properties, the astrophysical factor J of dSphs is usually quite high (Fig. 5.2) and, thanks to the studies of their stellar dynamics, more constrained than for some other classes of sources [19]. Still, due to the relatively small kinematic samples (especially for the ultra-faint dwarfs), it is not possible to determine whether their central dark matter distributions are cusped or cored. There are hints of dark matter cores in classical dSphs [20], but more definitive affirmation will have to wait for complete measurements of proper motions of the individual member stars, as well as higher-quality photometric data sets from the central regions.

As for the potential boost of the expected gamma-ray flux due to the presence of substructures, in the case of dSphs it is predicted to be negligible (factor of few at most, [21]), which leads to rather straightforward interpretation of the no-detection result as a limit on $\langle\sigma_{\rm ann}v\rangle$ or τ_χ (with respect to some other sources, like the Galactic Center or galaxy clusters, Sect. 2.3.2.1). Overall, dSph galaxies can be considered excellent targets for indirect dark matter searches. In the last couple of years, several of these satellites have been observed by the IACTs: Draco, Willman 1 and Segue 1 by MAGIC (all in mono mode, [23–25]), Draco, Ursa Minor, Bootes, Willman 1 and Segue 1 by VERITAS [26, 27], and Sagittarius, Carina, Canis Major and Sculptor by

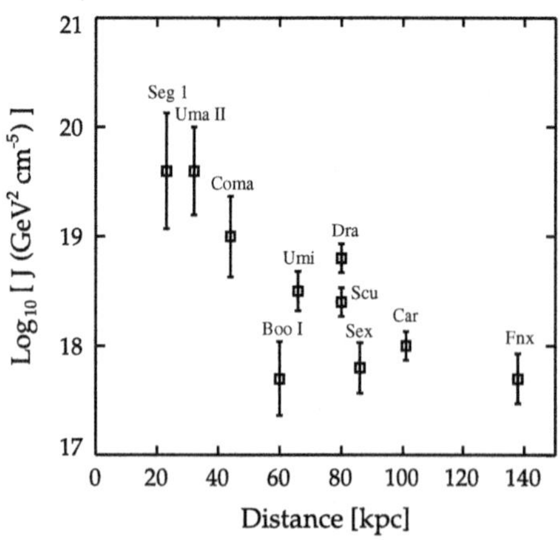

Fig. 5.2 Astrophysical factor J values for dSphs, integrated within a 0.5° radius, as a function of their galactocentric distances. The assumed dark matter density profile is NFW. Taken from [1]

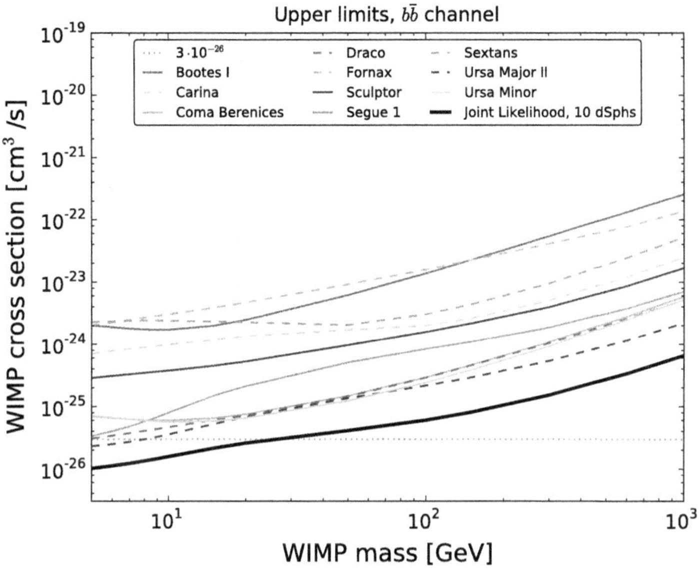

Fig. 5.3 Individual and combined limits on $\langle\sigma_{\text{ann}}v\rangle$ from 10 dSphs observed by Fermi-LAT, as a function of WIMP mass, assuming Br = 100 % annihilation to $b\bar{b}$. Taken from [22]

H.E.S.S. [28–30]. So far, no signal has been reported. Additionally, Fermi-LAT has combined the observations of 10 dSphs into the currently most constraining limits on $\langle\sigma_{\text{ann}}v\rangle$ in 5 GeV−1 TeV (Figs. 5.3 and 2.12, [22]).

5.1.3 Segue 1

This works sums up the 3-years long observational campaign of the MAGIC Telescopes directed to dark matter searches. The goal of this program was to perform deep survey of the best available (to MAGIC) dark matter source candidate, and accumulate many hours of good quality data. In the case of no detection, the sample is to be used for deriving limits on annihilation or decay rates for different dark matter models. The source chosen as the most suitable target is dSph galaxy Segue 1.

Segue 1 was discovered in 2006 as an overdensity of resolved stars in imaging data from SDSS. Located towards the Sagittarius constellation, at a distance of 23 ± 2 kpc, it was originally identified as a large globular cluster[1] [32], given its compactness and low absolute magnitude ($M_V = -1.5^{+0.6}_{-0.8}$), more characteristic for clusters than for dwarfs. This claim, however, was contested by the Keck/DEIMOS

[1]Thus the name Segue 1, after the survey SEGUE (Sloan Extension for Galaxy Understanding and Exploration, [31]), as conventional for globular clusters, instead after the constellation towards which it lies.

Fig. 5.4 Segue 1 as seen by the Keck II 10 m telescope. The image accounts for a total exposure of 5400 s in the 6400–9100 Å waveband. Segue 1 stars are circled in *green*. *Credit* M. Geha and W. M. Keck Observatory

(Deep Imaging Multi-Object Spectrograph, [33]) spectroscopic measurements of the radial velocities of 24 of the Segue 1 member stars: with mean heliocentric velocity estimated to ~206 km/s, and velocity dispersion of 4.2 ± 1.2 km/s, it was concluded that Segue 1 was actually an ultra-faint dwarf galaxy, and a highly dark matter dominated one [34]. There were several consequent attempts to dispute these results and classify the source as a globular cluster again (see, e.g. [35]), but, in 2011, a more comprehensive Keck/DEIMOS spectroscopic survey was conducted, increasing the number of member stars to 71 (Fig. 5.4). Analysis of these new data allowed for a clearer identification of Segue 1 as the least-luminous dSph galaxy, with the highest known M/L, estimated to ~3400 $M_{\odot}/L_{\odot}$ [16].

It should be clear that, despite this impressive M/L, its value has to be treated with caution. Although the mean velocity of Segue 1 member stars allows their clear separation from the foreground ones of the Milky Way, special care must be taken to make sure that no interloping star is included in the mass characterization of the system, as it can significantly inflate the total luminosity and therefore the M/L value. A more extensive future study of kinematic properties of the member stars should be able to address these uncertainties.

Some basic measured (or estimated) properties of Segue 1 are given in Table 5.2. As it can be seen, this dSph is relatively close, highly dark matter dominated and with almost no background of conventional origin at very high energies. In addition, it is located outside of the Galactic plane, it is in the Northern hemisphere and visible to MAGIC for about 370 h of dark time per year at low *Zd*. From the comparison with other potential dark matter sources, Segue 1 has been chosen as the (currently) best available target for indirect dark matter searches with MAGIC, and the following sections present the results of extensive observations of this dSph performed over the course of 3 years.

Table 5.2 Segue 1 main characteristics. Extracted from [16, 34]

Segue 1	
Coordinates	$10^{h}\ 07^{m}\ 04^{s}$, $+16°\ 04'\ 55''$
Distance	23 ± 2 kpc
Number of resolved stars	71
Magnitude	$-1.5^{+0.6}_{-0.8}$
Apparent magnitude	13.8 ± 0.5
Luminosity	$340\ L_{\odot}$
Mass	$5.8^{+8.2}_{-3.1} \times 10^{5}\ M_{\odot}$
M/L	$\sim 3400\ M_{\odot}/L_{\odot}$
Half-light radius	29^{+8}_{-5} pc
System velocity	208.5 ± 0.9 km/s
Velocity dispersion	$3.7^{+1.4}_{-1.1}$ km/s
Mean [Fe/H]	-2.5

5.2 Observations and Data Reduction

Observations of Segue 1 were performed with MAGIC, in stereoscopic mode, between January 2011 and February 2013. During this time, the telescopes underwent a series of significant changes: at the end of 2011, the readout systems of both instruments were replaced by the more advanced, DRS4-based configurations (Sect. 3.1.3.2); by the end of 2012, among other improvements, the camera of MAGIC-I was upgraded to the exact replica of that of MAGIC-II (Sect. 3.1.2). As a result, the performance of the system varied during the total period of Segue 1 observations; therefore, data from each of the different telescopes states are processed separately, before being combined through the full likelihood into the final results (Eq. (4.5), Sect. 5.4). Basic details are summarized in Table 5.3.

5.2.1 Sample A: 2011 Data

Data sample A refers to the Segue 1 observations from January to May 2011, carried under dark night conditions, for total of 64 h. The *Zd* range extended from 13° (corresponding to the source culmination at MAGIC latitude) to 33.7°, thus ensuring the low energy threshold of the analysis.

Observational Setup Observations were performed in a false tracking (wobble) mode (Sect. 3.2.1); however, instead of the standard observational setup (two pointing positions at opposite directions and a 0.4° distance from the source), a special wobble configuration was devised for this case. Namely, there is a 3.5-apparent magnitude star, Eta Leonis (η Leo), located at a 0.68° angular distance from the nominal position

Table 5.3 Basic observational and data reduction details for four samples of Segue 1 data

	Sample A	Sample B1	Sample B2	Sample C
Readout	DRS2	DRS4	DRS4	DRS4
MAGIC-I camera	Old	Old	Old	New
Obs. period	Jan–May 2011	Jan–Feb 2012	Mar–May 2012	Nov 2012–Feb 2013
Obs. time [h]	64	24.28	59.77	55.05
Zd range [deg]	13–33.7	13–32.5	13–35.7	13–37
Az range [deg]	104.8–250.2	120.2–252.0	115.4–257.2	103.8–259.4
Wobble around	SegueJ	SegueJ	SegueJ	Segue 1
Wobble offset [deg]	0.29	0.29	0.29	0.40
Wobble angles [deg]	5,185	5,185	5,185	5,185
Image cleaning	Absolute	Absolute	Absolute	Dynamic sum
q_c/q_b [ph.e.] (M-I)	6/3	6/3	6/3	6/3.5
q_c/q_b [ph.e.] (M-II)	9/4.5	8/4	8/4	6/3.5
Δt_c [ns]	4.5	4.5	4.5	4.5
Δt_b [ns]	1.5	1.5	1.5	1.5
W1 $t_{\rm eff}$ [h]	22.66	6.07	25.02	23.71
W2 $t_{\rm eff}$ [h]	24.35	6.20	26.11	23.80
$t_{\rm eff}$ [h]	47.00	12.26	51.13	47.51
Total $t_{\rm eff}$ [h]				**157.9**

Refer to the main text for additional explanations

of Segue 1, and as such, in the FoV of MAGIC cameras. Every time the star enters the trigger area of the cameras (Fig. 3.5), it causes an increase in the rate of the corresponding PMTs and in the rate of data acquisition. To solve this problem, the trigger threshold of the affected pixels has to be dynamically augmented during the observations each night (since the star moves in the camera). This, however, in particular was an issue for MAGIC-I, as the automatic pixel rate control of its camera was not fast enough to quickly correct for the presence of η Leo.

In order to deal with the problem, the wobbling scheme was chosen in such a way that the star was always outside the trigger region of MAGIC-I: instead of wobbling around the nominal position of Segue 1, tracking was done with respect to a 'dummy' position, dubbed SegueJ, located 0.27° away from Segue 1 and on the opposite side from η Leo (Fig. 5.5). The wobble offset of 0.29° and wobble angles of 175° and 355° were chosen so that the star was always at a 1° distance from the camera center and therefore outside of the MAGIC-I trigger region. Also, with such configuration, Segue 1 is 0.4° away from the camera center (as in the standard observational configuration) with residual background estimated from positions 0.58° away. That way, the homogeneity of the camera exposure of Segue 1 and *OFF* regions, as well as the sensitivity of the analysis were preserved.

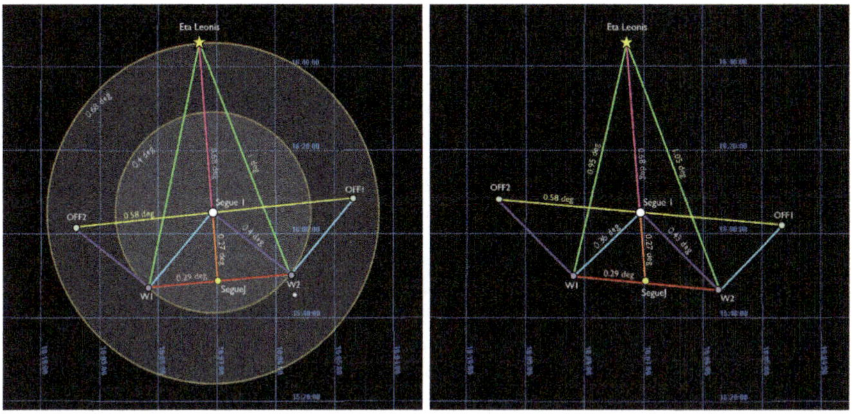

Fig. 5.5 Observational scheme devised to exclude the η Leo star from the MAGIC-I trigger area (*left*) differs from the one actually used in the observations (*right*), due to different conventions applied in coordinate calculations. Nevertheless, the sensitivity was not affected. This configuration was used for the Segue 1 observations from January 2011 to May 2012 (samples A, B1 and B2). See the main text for more details

Unfortunately, due to different conventions (regarding the direction of right ascension increase) applied in the calculations of wobble scheme by the analyzers and by the Central Control of the telescopes, instead of 175° and 355° wobble angles, 185° and 5° were used. This altered the desired configuration, with angular distances between the wobble positions and Segue 1 being 0.36° and 0.43° instead of 0.4° (Fig. 5.5). These differences, however, are not sufficiently significant to affect the sensitivity of the analysis (for instance, mispointing of the telescopes can be as big as 0.1°). Furthermore, the background estimation is done from positions that are at the same distances from the camera center as the source, and the distances between these *OFF* positions and Segue 1 are the same, and remain 0.58°; thus, the exposure of *ON* and *OFF* regions is homogeneous.

In order to avoid possible effects of the 'stereo blob' (Sect. 3.2.1), special care was taken to ensure that both wobble positions were observed for similar amounts of time and with similar azimuthal (*Az*) coverage (Fig. 5.6). For this, observations were reviewed on daily basis, faulty subruns rejected, and estimates made on how the new data (to be taken the following night) would contribute to the existing *Az* distribution.

Data Reduction Data reduction was performed following the steps of the standard MAGIC procedures described in Sect. 3.3. After the calibration (Sect. 3.3.1), the absolute image cleaning was applied using the standard values for the pre-upgrade configuration (Sect. 3.3.2): for MAGIC-I, a minimum of $q_c = 6$ ph.e. and $q_b = 3$ ph.e. charge for core/boundary pixels was required, while for MAGIC-II those numbers were $q_c = 9$ ph.e. and $q_b = 4.5$ ph.e. At the same time, individual core pixels whose signal arrival time differed from the mean core time by more than $\Delta t_c = 4.5$ ns were rejected, as well as the individual boundary pixels with arrival time difference

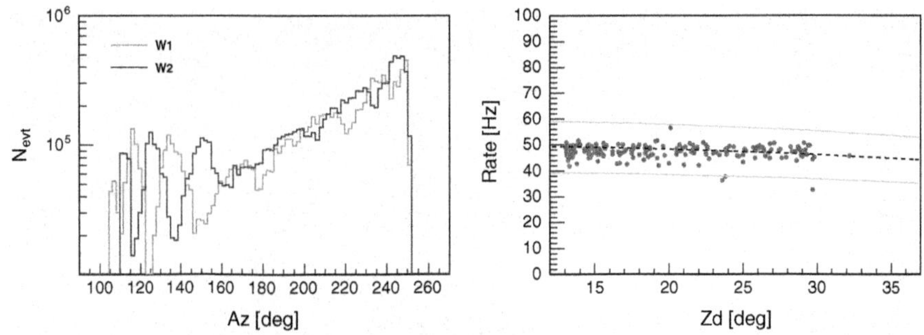

Fig. 5.6 *Left Az* distribution for the sample A observations for both wobble positions. *Right* Distribution of rates for the same period as a function of *Zd*. *Dashed line* represents the reference rate and full lines its ±20 % values, marking the area outside of which the data are excluded. Each point represents a single data run

with respect to that of their core neighbors greater than $\Delta t_c = 1.5$ ns. Cleaning was followed by the image parametrization.

Next, quality selection was performed (Sect. 3.3.3), first individually for data of each telescope and each observational day. Criteria for rejection of a given subrun was deviation for more than 20 % with respect to average (for the day in question) value of the event rate, and 5 % deviation from the mean of *length* and *width* parameter distributions. This way, 12.3 h of data were excluded, mainly due to the bad meteorological conditions and subsystem failures. After this individual quality selection, data from both instruments were merged and stereo parameters reconstructed (Sect. 3.3.4). This was followed by another quality check, but this time encompassing the entire sample: runs whose event rates differed by more than 20 % from the mean rate of the whole data set were excluded from the further analysis (Fig. 5.6).

The event characterization—γ/hadron separation, energy and arrival direction estimation—was performed following the procedures outlined in Sect. 3.3.5. The *hadronness* calculation (Sect. 3.3.5.1) was done by means of the RF method; the corresponding matrices were trained over a subset of MC gamma-ray events and ~16 h-sample composed of observations where no significant gamma-ray emission was found. These datasets were chosen because they share (or are simulated with) similar observational conditions as Segue 1 (*Zd*, instrument PSF). The energy estimation (Sect. 3.3.5.3) was derived from LUTs constructed from MC gamma-ray events. As for the arrival direction, it was calculated with the *Disp RF* method (Sect. 3.3.5.2), trained using the image and shower parameters, as well as the *Zd* of observations.

The validity of the created RF matrices and energy LUTs was first verified on the 4.8 h of Crab Nebula data, contemporaneous to the sample A observations and of same *Zd* range, before being used in the characterization of Segue 1 events: the obtained integral sensitivity of 0.79 ± 0.10 % C.U. above 250 GeV agrees with the expectations (Fig. 3.13). The same Crab Nebula sample was later used for optimization of the analysis cuts (Sect. 5.3.1).

Finally, after the correction for dead time of the system (0.5 ms), t_{eff} of the Segue 1 A sample was reduced to 47.0 h of good-quality data.

5.2.2 *Sample B: 2012 Data*

Sample B refers to the Segue 1 observations with MAGIC between January and May 2012. Since December of 2011, telescopes were operating with improved, DRS4-based readout systems, which reduced the dead time from 0.5 ms to only 26 μs. However, the beginning of the post-upgrade period was marked by many hardware problems and analysis issues. In particular, the Segue 1 data taken during January and February were affected by several faulty mezzanines in MAGIC-I (total of 24 bad channels); furthermore, due to a bug in the AMC system of MAGIC-I, the PSF of this telescope was increased. A posteriori, data taken under such conditions could be recovered through software corrections, but they needed special attention and tailored MC files. The issues in question were solved for observations from March onwards; however, as the treatment and MC sets for these data are different, sample B is divided into two separately analyzed sets: sample B1 (pre-March) and sample B2 (post-March data).

5.2.2.1 Sample B1: Pre-March Data

These data were taken from the end of January to the end of February 2012, for total of 24.3 h, under dark night conditions and for *Zd* range between 13° and 32.5°.

Observational Setup Observations were performed in wobble mode, using the same configuration as for the sample A (Sect. 5.2.1, Fig. 5.5). Due to the very harsh weather conditions in this period, and a rather unstable system, data taking and balancing of azimuthal coverages for both wobbles was complicated (Fig. 5.7).

Data Reduction Data reduction followed the standard steps described in Sect. 3.3. After the calibration (from this period on done by the program `sorcerer`), absolute image cleaning was applied, with the cleaning parameters specified in Table 5.3. Following the same data selection criteria as described for the A sample, almost half of these observations (~12 h, Fig. 5.7) were rejected, which was not surprising given the very bad weather and frequent system failures during this period. Processes of the event reconstruction were done with matrices and LUTs trained with ~2 h of data where no gamma-ray signal was found, and with MC sample simulated so that hardware issues of MAGIC-I (the increased PSF and broken channels in the camera) were taken into the account. Validity of reconstruction matrices and tables was verified on ~4 h of contemporaneous (to Segue 1) Crab Nebula sample: above 250 GeV, the integral sensitivity was calculated to be $0.81 \pm 0.08\,\%$ C.U. This sample was later used in the analysis for the cuts optimization (Sect. 5.3.1).

After the dead time correction, t_{eff} of the Segue 1 B1 sample amounted to 12.26 h of good-quality data.

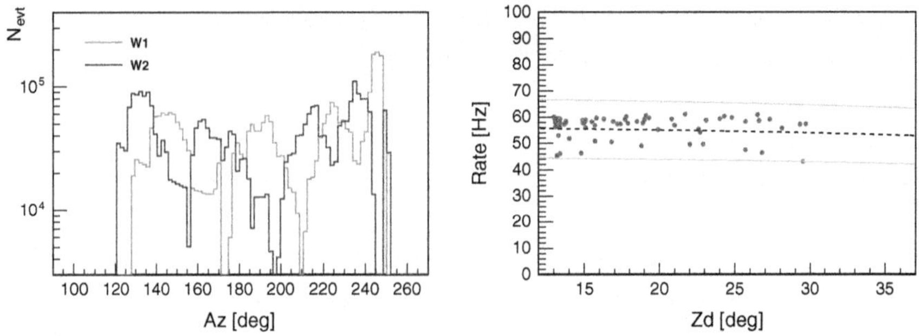

Fig. 5.7 *Left Az* distribution for the sample B1 observations for both wobble positions. *Right* Distribution of rates for the same period as a function of *Zd*. *Dashed line* represents the reference rate and full lines its ±20 % values, marking the area outside of which the data are excluded. Each point represents a single data run

5.2.2.2 Sample B2: Post-March Data

Between March and May 2012, Segue 1 was observed for 59.8 h, under dark night conditions and for *Zd* range from 13° to 35.7°, thus ensuring a low energy threshold.

Observational Setup Same wobble configuration as for the samples A and B1 was used (Sect. 5.2.1, Fig. 5.5). Azimuthal coverage of each of the wobble positions was carefully controlled during the data taking (Fig. 5.8).

Data Reduction After the calibration, absolute image cleaning was applied (the cleaning parameter values are given in Table 5.3). After the data selection, 6.9 h of observations were rejected, mainly due to the subsystem failures during the data taking and unstable atmospheric conditions (Fig. 5.8). Matrices and LUTs for event characterization were obtained from the corresponding MC sample and from ∼6 h of actual data that contained no signal. Out of this 6 h, ∼1.5 h are of Segue 1 itself: this measure was necessary since there were no data from other sources, taken in

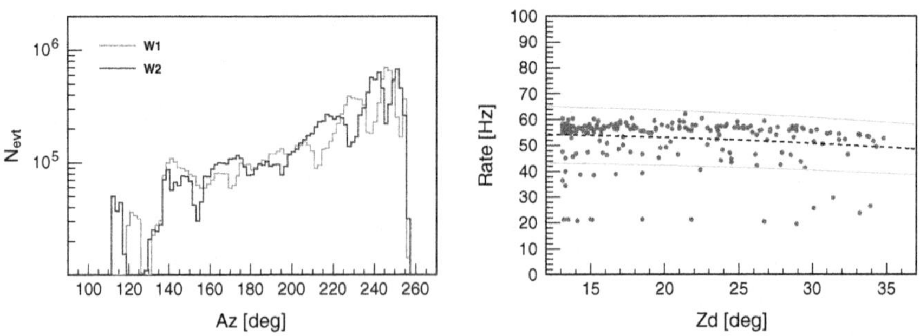

Fig. 5.8 *Left Az* distribution for the sample B2 observations for both wobble positions. *Right* Distribution of rates for the same period as a function of *Zd*. *Dashed line* represents the reference rate and full lines its ±20 % values, marking the area outside of which the data are excluded. Each point represents a single data run

the same period and under the same conditions as Segue 1 that, after the quality checks, covered the *Zd* range between 17° and 28°. To make sure that matrices for these angles were properly calculated, a fraction of Segue 1 sample, overlapping the mentioned range, was added for the training. Those data were not considered in the further analysis. Generated LUTs and matrices were first tested on 6.2 h of Crab Nebula sample, allowing for a $0.79 \pm 0.07\,\%$ C.U. integral sensitivity above 250 GeV, before being applied in event reconstruction of the Segue 1 data. The Crab Nebula sample was also used for the optimization of the analysis cuts later on (Sect. 5.3.1).

After the described quality selection, dead time correction and removal of ~1.5 h of data used for RF training, t_{eff} of the Segue 1 B2 sample was 51.13 h of good-quality data.

5.2.3 Sample C: 2012-2013 Data

By November 2012, the planned improvements of the MAGIC Telescopes were successfully completed. Most notably, the camera of MAGIC-I was replaced with an exact replica of the MAGIC-II camera (Sect. 3.1.2). The changes made in the upgrade were followed by some changes in the analysis chain and MC production.

The sample C refers to the observations of Segue 1 from November 2012 to February 2013. For total of 55 h, under dark night conditions, Segue 1 was observed in the $13^\circ - 37^\circ$ *Zd* range.

Observational Setup The upgraded MAGIC-I camera has greater trigger area and faster electronics than the old one. In particular case of the Segue 1 observations, this means that the η Leo is always inside the trigger region (as in the case of MAGIC-II), but also that the camera electronics are now capable of a swift correction of the trigger thresholds for the PMTs affected by the star. It was, therefore, opted to observe Segue 1 with the standard wobble settings: 0.4° angular distance between the source and camera center and 0.8° between the source and *OFF* regions (in both wobble positions, Fig. 5.9). The offset angles remained 5° and 185°, in order to keep the symmetry between the positions of the source, pointings and η Leo at maximum. The azimuth coverage of both wobble positions was again carefully monitored during the course of observations (Fig. 5.10).

Data Reduction Since 2013, the standard image cleaning in the data reduction chain is the dynamical sum cleaning (Sect. 3.3.2)—that is, the cleaning levels are adapted for each event depending on the *size* of the shower, for the core pixels with charge above the basic threshold of 750 ph.e. For this sample, the default cleaning level values were used (Sect. 3.3.2, Table 5.3). As a result of data selection (performed in the same way as for the other samples), 6.65 h of observations were rejected from the sample C. For event characterization, training was done using a MC sample and ~10 h of observations of sources where no significant gamma-ray excess was found. Validity of the generated matrices and LUTs was verified on ~30 h of contemporaneous (to Segue 1) Crab Nebula data, and an integral sensitivity of $0.70 \pm 0.03\,\%$

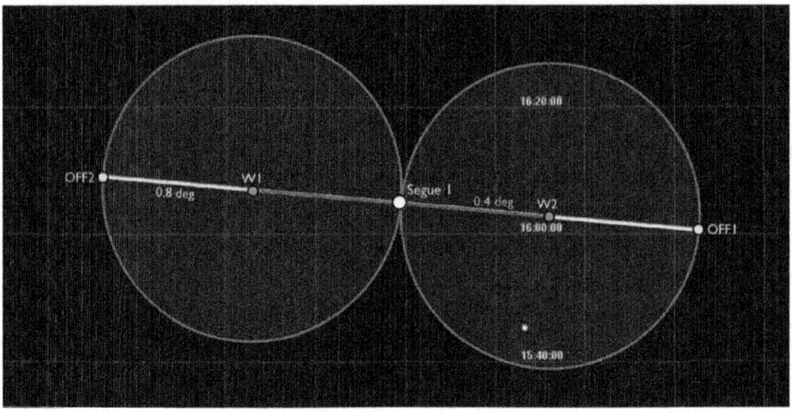

Fig. 5.9 Observational scheme for the sample C observations. The wobbling is done around Segue 1, at standard offset of 0.4° and with wobble angles of 5° and 185°, preserving the maximal symmetry between the source, pointing and η Leo. See the main text for more details

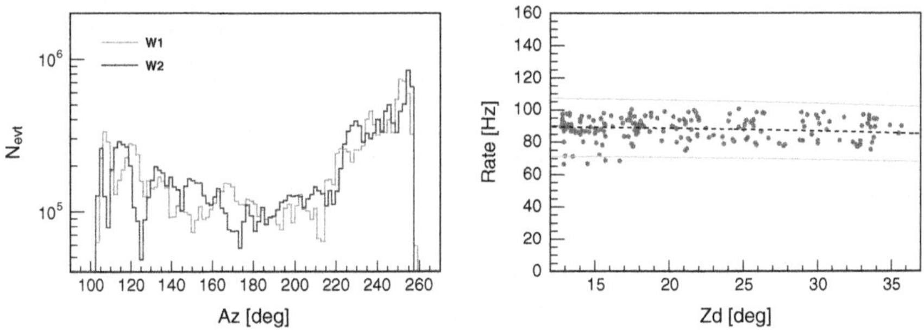

Fig. 5.10 *Left Az* distribution for the sample C observations for both wobble positions. *Right* Distribution of rates for the same period as a function of *Zd*. *Dashed line* represents the reference rate and full lines its ±20 % values, marking the area outside of which the data are excluded. Each point represents a single data run

was obtained above 250 GeV. The Crab Nebula sample was used later on for the cuts optimization (Sect. 5.3.1).

After the dead time correction, $t_{\rm eff}$ of the Segue 1 sample C was reduced to 47.51 h of good-quality data.

5.3 Analysis

This section describes the full likelihood analysis of the Segue 1 data. Following the prescription of this method, presented in Sect. 4.2.1, the response function of the telescopes is determined, the background is modeled, and the spectral function depicting the expected signal distribution is decided upon (by choosing a certain dark matter model). But first, the analysis cuts optimal for a source such as Segue 1 are defined. In addition, for the sake of completeness, as well as for consistency tests and cross-check purposes, the high-level products of the MAGIC standard analysis chain (Sect. 3.3.7) are shown; however, those results will not be used at the end of this chapter for the interpretation of the Segue 1 data in the context of dark matter searches. Lastly, the calculation of the astrophysical factor J for Segue 1 is presented at the end of this Section.

5.3.1 Cuts Optimization

The choice of the cuts to be applied in the analysis is very important, as it directly influences the prospects for detection. Segue 1 is assumed to be a weak source, therefore the cuts for this analysis are rather tight, selected to ensure the best possible sensitivity.

The main cuts applied in this work are:

- *size*: given how it is in the interest of this analysis to cover the low energy range, the *size* cut (for both telescopes) requires minimum of only 50 ph.e. for an event not to be rejected;
- *hadronness*: maximal exclusion of the background implies low hadronness. This cut is optimized on a Crab Nebula sample, assuming a weak source such as Segue 1;
- θ^2: radius of *ON* and *OFF* regions (given as θ) defines how big are the areas from which the events are considered in the analysis. It is also optimized on the Crab Nebula sample;
- E_{min} and E_{max}: as shown in Sect. 4.2.2, the full likelihood method takes advantage of the entire energy range available. Therefore, the lower and upper cuts in energy for this analysis are set to $E_{min} = 59.2\,\mathrm{GeV}$ and $E_{max} = 10\,\mathrm{TeV}$, respectively. The value for E_{min} does not go down to even lower energies since the response of the detector becomes unreliable; E_{max} does not exceed 10 TeV as there are no remaining events above this limit (after the *hadronness* and θ^2 cuts are applied).

As already mentioned in the previous section, the Crab Nebula data sample is used for the optimization of the analysis cuts (of a given observational period). In this work, this is done in the following way: θ^2 and *hadronness* are varied within a 2-dimensional grid (adopting values from 0.01 to 0.05 deg^2 for the θ^2 cut, and from 0.1 to 0.5 for *hadronness*). Each pair of cuts on the grid is applied (together with the fixed energy and *size* cuts) to the Crab Nebula sample, and the surviving signal and background events (N_{ON} and N_{OFF}, respectively) are used to calculate the significance following the Li&Ma prescription (Sect. 3.3.6, Eq. (3.5)). However, as Segue 1 is expected to be a much weaker source than the Crab Nebula, its signal is assumed to be only 5 % of that of the Crab, thus $N_{ex} = (N_{ON} - N_{OFF})$ is scaled down to 5 %. The resulting significances for different sets of cuts from θ^2-*hadronness*

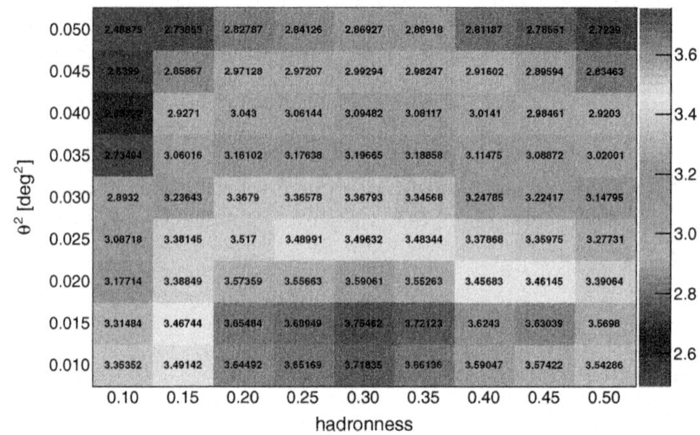

Fig. 5.11 Significance scan over the different values of θ^2 (y-axis) and *hadronness* cuts (x-axis), for *size* > 50 ph.e. and energy between 59.2 GeV and 10 TeV. The *color scale* refers to the significance, obtained after the cuts are applied, for a weak source with an assumed flux that is 5 % of the Crab Nebula flux. This particular case refers to the sample A. See the main text for more details. (Color in online)

plane are compared, and the cuts corresponding to the highest value are chosen as optimal (Fig. 5.11).

Given that both θ^2 and *hadronness* cuts are energy-dependent, the correct course would be to optimize these cuts for different energy bins. Indeed, this was originally done for the analysis presented here. However, the binned approach led to distributions of events from *ON* and *OFF* regions that were step-like, with 'steps' appearing at bin edges. Such distributions are likely to cause practical problems in the computation of the full likelihood, as 'step' could be mistaken for a feature. Therefore, it was opted to go for a global set of cuts, optimized for the entire energy range. The sensitivity was not affected by this in a significant way: it has been verified that, for the Crab Nebula sample, when both the optimal and global cuts are applied for each energy bin, the resulting sensitivities and significances are compatible—the sensitivity differences are, for all four observational periods, within the statistical uncertainties.

The complete list of analysis cuts, optimized for each of the four observational periods separately, is given in Table 5.4. Aside from being used on the Segue 1 data, the same cuts are also applied to the MC gamma-ray events for the calculations of the response functions (Sect. 5.3.3).

5.3.2 Results of the Standard Analysis

As already explained in Sect. 4.1.1, the conventional analysis approach—currently the standard for the IACTs—is not optimal for signals with some characteristic

Table 5.4 Optimized analysis cuts for different Segue 1 observational periods

Cuts	Sample A	Sample B1	Sample B2	Sample C
$E_{min} - E_{max}$ [GeV]	59.2 – 10000.0	59.2 – 10000.0	59.2 – 10000.0	59.2 – 10000.0
Size [ph.e.]	50	50	50	50
Hadronness	0.30	0.30	0.25	0.25
θ^2 [deg^2]	0.015	0.015	0.015	0.015

See the main text for more details

features, as expected from dark matter annihilation or decay (Sect. 2.3.2.3). Thus, in this work, the full likelihood method is used instead. Still, for the sake of completeness and comparison with different analyses and instruments, some of the high level products of the standard MAGIC analysis chain are presented (Sect. 3.3.7).

Figure 5.12 shows the θ^2-plot for the whole Segue 1 sample. The distributions of events from *ON* (red) and *OFF* regions (gray) are first calculated for each wobble position of each observational period (with the corresponding cuts, Fig. 5.13), before being combined into the final result. As it can be seen, no significant gamma-ray signal is found. It should be noted that there is a negative excess in the distribution at low θ^2 cut values; its magnitude, however, is only $\sim$1.5 % of the residual background, well within the systematic uncertainties in the background estimation at the lowest energies ($\sim$5 %).

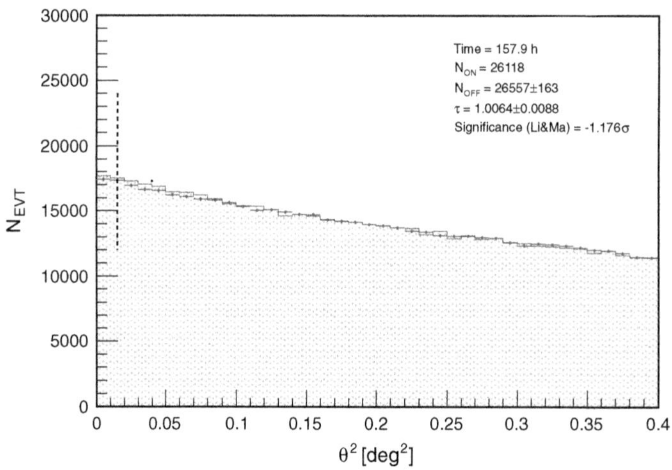

Fig. 5.12 Segue 1 cumulative θ^2 distribution from 157.9 h of observations. The signal (*ON* region) is presented by *red points*, while the background (*OFF* region) is the *shaded gray area*. The *OFF* sample is normalized to the *ON* sample in the region where no signal is expected, for θ^2 between 0.15 and 0.4 deg^2. The vertical dashed line shows the θ^2 cut

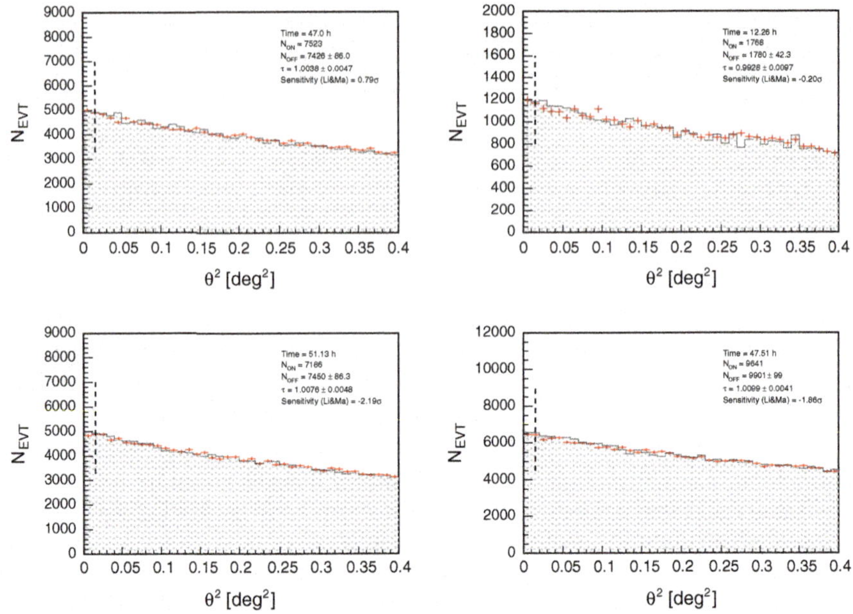

Fig. 5.13 Segue 1 θ^2 distributions for four observational samples, each with its own cuts applied: sample A (*top left*), sample B1 (*top right*), sample B2 (*bottom left*) and sample C (*bottom right*). The color coding is the same as in Fig. 5.12

Figure 5.14 is the total skymap for the Segue 1 observations. Again, each of the four periods (A, B1, B2 and C) are first mapped with their own cuts before being combined. The lower energy cut is (for all samples) set to $E_{\rm min} = 100\,{\rm GeV}$, to avoid some known difficulties related to the skymap reliability at the lowest energies. As seen from the Fig. 5.14, there is no notable excess at the Segue 1 position. The most significantly positive fluctuations seen in the map appear in the neighborhood of the bright star η Leo, north of Segue 1. Those cannot be attributed to a real gamma-ray signal, but more likely to inefficiencies in the image cleaning for the pixels close to the star, due to the signals induced by its direct light.

Since these results agree with the no-signal-over-the-background scenario, differential and integral flux upper limits are calculated for the gamma-ray emission from the source, assuming a power law-shaped spectra with slope Γ, and by relying on the conventional approach in the analysis.

The differential flux upper limits are estimated in energy bins ΔE as:

$$\frac{d\Phi^{\rm UL}}{dE}(E'_\star) = \frac{N_{\rm ex}^{\rm UL}(\Delta E)}{t_{\rm eff}} \frac{1}{\int_0^\infty \; A_{\rm eff}(E'; \Delta E) S(E') dE'}, \tag{5.1}$$

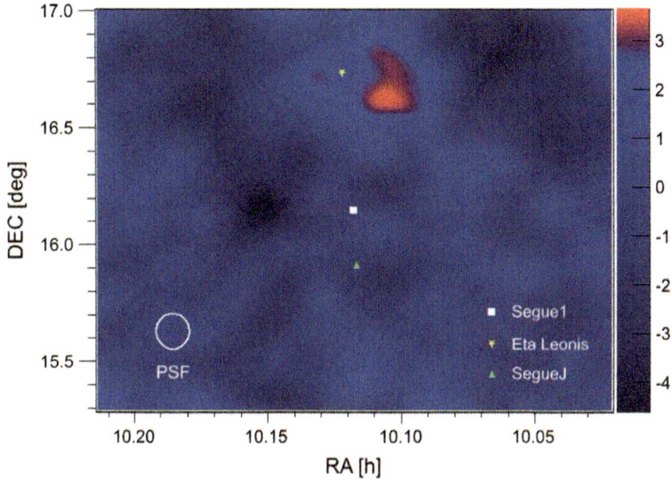

Fig. 5.14 Segue 1 cumulative significance skymap, for 157.9 h of observations and energies above 100 GeV. Positions of Segue 1, η Leo and SegueJ are shown as *white square*, *yellow star* and *green triangle*, respectively

where $S(E') = (E'/E'_\star)^\Gamma$ is the chosen power law spectrum. $E'_\star$ is the pivot energy for the particular energy bin, defined as:

$$E'_\star = \frac{\int_0^\infty E' S(E') A_{\rm eff}(E'; \Delta E) dE'}{\int_0^\infty S(E') A_{\rm eff}(E'; \Delta E) dE'}. \quad (5.2)$$

$A_{\rm eff}(E'; \Delta E)$ is the effective area within ΔE, for gamma-rays of true energy E'; $N_{\rm ex}^{\rm UL}$ is computed using global $N_{\rm ON}$, $N_{\rm OFF}$ and τ values and the method of Rolke (Sect. 3.3.7.5, [36]), for a 95 % c.l. and assuming a systematic uncertainty on the overall detection efficiency of 30 %. $A_{\rm eff}(E'; \Delta E)$ is computed for the entire sample as the weighted average of the effective areas of the four considered data sets, with weights being the corresponding observation times, i.e.:

$$A_{\rm eff}(E'; \Delta E) = \frac{\sum_i \left(A_{{\rm eff}\,i}\, t_{{\rm eff}\,i}\right)}{t_{\rm eff}}, \quad (5.3)$$

where the index i runs over the different samples.

Table 5.5 summarizes the upper limits in four (estimated) energy logarithmic bins, between 100 GeV and 10 TeV, for different assumed slopes of the power law-shaped signal emission ($\Gamma = -1.0, -1.5, -1.8, -2.0, -2.2$ and -2.4). The results are also shown on Fig. 5.15.

The integral flux upper limits are calculated above a certain energy threshold $E_{\rm th}$, based on the prescription from Eq. (3.12). The values of $N_{\rm ex}^{\rm UL}(E > E_{\rm th})$ and $A_{\rm eff}(E'; E > E_{\rm th})$ are obtained in a similar manner as described for the differential

Table 5.5 The differential flux upper limits from 157.9h of Segue 1 observations with MAGIC, in four energy bins and for several power law-shaped spectra

ΔE [GeV]	N_{ON}	N_{OFF}	$N_{\mathrm{ex}}^{\mathrm{UL}}$	$d\Phi^{\mathrm{UL}}/dE$ [TeV^{-1} cm^{-2} s^{-1}]					
				$\Gamma = -1.0$	$\Gamma = -1.5$	$\Gamma = -1.8$	$\Gamma = -2.0$	$\Gamma = -2.2$	$\Gamma = -2.4$
100–320	11135	11042	653.98	1.29×10^{-11}	1.37×10^{-11}	1.45×10^{-11}	1.51×10^{-11}	1.59×10^{-11}	1.69×10^{-11}
320–1000	1606	1608	81.44	2.38×10^{-13}	2.53×10^{-13}	2.67×10^{-13}	2.78×10^{-13}	2.92×10^{-13}	3.07×10^{-13}
1000–3200	341	314	108.93	6.44×10^{-14}	6.99×10^{-14}	7.48×10^{-14}	7.88×10^{-14}	8.36×10^{-14}	8.96×10^{-14}
3200–10000	59	55	33.33	3.91×10^{-15}	4.34×10^{-15}	4.79×10^{-15}	5.19×10^{-15}	5.71×10^{-15}	6.48×10^{-15}

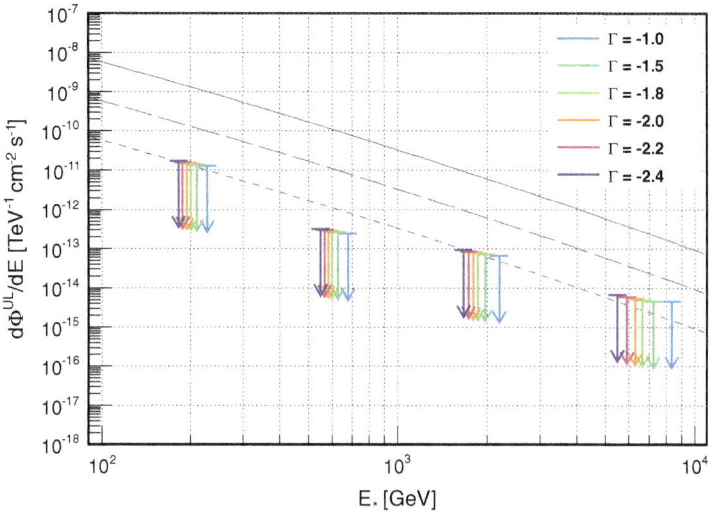

Fig. 5.15 The differential flux upper limits from 157.9 h of Segue 1 observations with MAGIC, assuming a power law-shaped signal emission and different spectral slopes Γ. As a reference, the Crab Nebula differential flux (*full line*, [37]) and its 10 % and 1 % fractions (*long-dashed* and *dashed lines*, respectively), are also drawn

upper limits. Again, several power law spectra are assumed, with different slopes Γ. The results are summarized in Table 5.6 and also shown in Fig. 5.16. As it can be seen, the integral upper limits are affected by statistical fluctuations originated by the positive or negative excess in the number of events above E_{th}. In order to estimate the effect Γ and E_{th} have on the Φ^{UL} value, without being biased by these statistical fluctuations, Fig. 5.16 also shows the integral upper limits computed assuming that $N_{\text{ex}}(E > E_{\text{th}}) = 0$.

5.3.3 Response Function

In the context of this work, the response function is fully defined by the energy resolution σ, the energy bias μ and the effective area A_{eff} of the telescopes, all as a function of true energy. For each of the four observational periods discussed in Sect. 5.2, the performance of the system is evaluated separately, using the corresponding MC *test* samples (Sect. 3.3.5). Each *test* MC set is first subjected to the processes of event characterization, using the same matrices and LUTs as for the Segue 1 data, after which the cuts optimized for the source are applied (Table 5.4).

The energy reconstruction functions, σ and μ are obtained in the following way: the MC gamma-ray events that survive the cuts are divided in a number of the (true) energy bins. For each bin, distribution $(E - E')/E'$ is fitted with a Gaussian function

Table 5.6 The integral flux upper limits from 157.9 h of Segue 1 observations with MAGIC, for different energy thresholds and several power law-shaped spectra

E_{th} [GeV]	N_{ON}	N_{OFF}	N_{ex}^{UL}	Φ^{UL} [cm^{-2} s^{-1}]					
				$\Gamma = -1.0$	$\Gamma = -1.5$	$\Gamma = -1.8$	$\Gamma = -2.0$	$\Gamma = -2.2$	$\Gamma = -2.4$
59.2	26118	26557	196.67	4.13×10^{-13}	6.84×10^{-13}	9.48×10^{-13}	1.16×10^{-12}	1.38×10^{-12}	1.62×10^{-12}
100.0	13145	13023	66.57	1.28×10^{-13}	1.80×10^{-13}	2.22×10^{-13}	2.52×10^{-13}	2.80×10^{-13}	3.07×10^{-13}
177.8	4869	4913	139.15	2.59×10^{-13}	3.04×10^{-13}	3.47×10^{-13}	3.75×10^{-13}	4.01×10^{-13}	4.24×10^{-13}
317.2	2034	2007	141.99	2.33×10^{-13}	2.63×10^{-13}	2.83×10^{-13}	2.96×10^{-13}	3.08×10^{-13}	3.18×10^{-13}
562.3	878	855	123.51	1.90×10^{-13}	2.02×10^{-13}	2.10×10^{-13}	2.14×10^{-13}	2.18×10^{-13}	2.21×10^{-13}
1000.0	404	373	127.49	1.89×10^{-13}	1.93×10^{-13}	1.96×10^{-13}	1.98×10^{-13}	1.99×10^{-13}	1.99×10^{-13}
1778.3	175	173	56.04	8.17×10^{-14}	8.23×10^{-14}	8.24×10^{-14}	8.23×10^{-14}	8.19×10^{-14}	8.09×10^{-14}
3172.3	64	60	37.06	5.44×10^{-14}	5.43×10^{-14}	5.39×10^{-14}	5.35×10^{-14}	5.28×10^{-14}	5.12×10^{-14}
5623.4	20	19	20.60	3.23×10^{-14}	3.23×10^{-14}	3.14×10^{-14}	3.08×10^{-14}	2.99×10^{-14}	2.80×10^{-14}

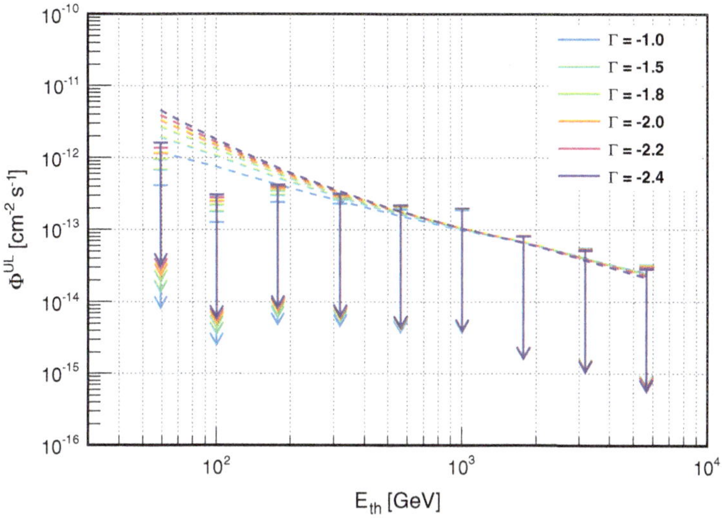

Fig. 5.16 The integral flux upper limits from 157.9 h of Segue 1 observations with MAGIC, assuming a power law-shaped signal emission and different spectral slopes Γ. *Dashed lines* indicate the integral upper limits if a zero significance is assumed

(Fig. 5.17). The width of that fit is taken as σ, and the mean of the fit as μ, for the energy range covered by the given bin. The final distributions of σ and μ as a function of E' for the four observational periods are shown in Fig. 5.18.

As for A_{eff}, it is obtained following Eq. (3.7). Given its dependence on the *Zd*, for the particular case of Segue 1 observations, A_{eff} distributions are first calculated for two bins in zenith: $Zd_1 = 13.0° - 25.8°$ and $Zd_2 = 25.8° - 35°$, which are then averaged, using the relative observation times of each *Zd* range as weights, into a global A_{eff} actually used in the analysis (Eq. (3.8)). Figure 5.18 shows A_{eff} distributions for the four considered observational periods, as a function of E'. Also plotted are the trigger effective areas of each period (i.e. A_{eff} before any of the cuts are applied),

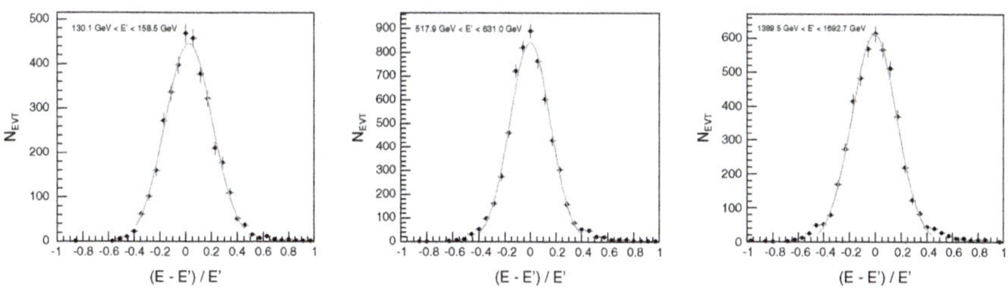

Fig. 5.17 Estimation of the energy resolution σ and bias μ in different bins of true energy E': the $(E - E')/E'$ distribution is fitted with a Gaussian, with σ being the width of the fit and μ the mean of the fit

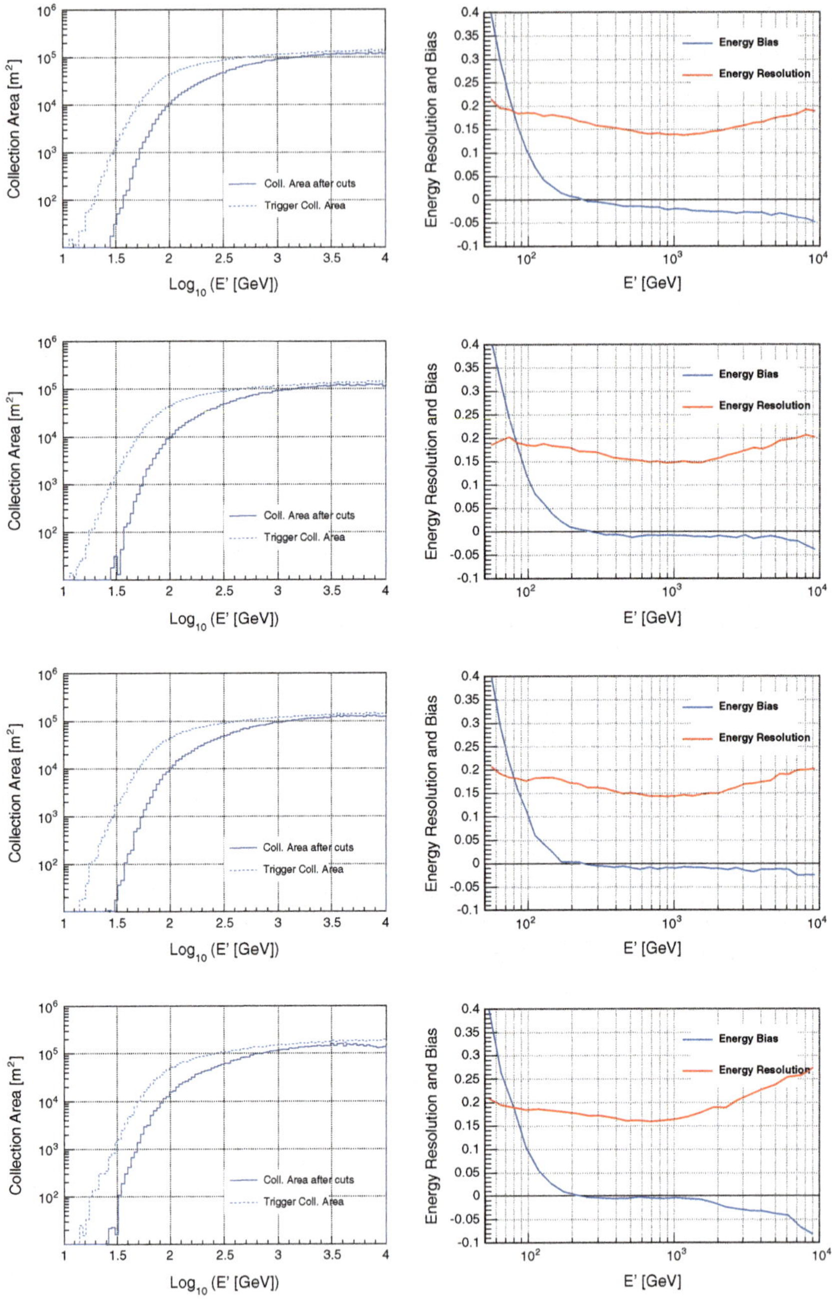

Fig. 5.18 Effective area A_{eff} (*left column*) and energy resolution σ and bias μ (*right column*) for different periods of Segue 1 observations: sample A (*first row*), sample B1 (*second row*), sample B2 (*third row*) and sample C (*fourth row*)

illustrating how big was (is) the effective surface around the telescopes within which a photon of certain energy would have to fall in order to trigger the system.

5.3.4 Background Modeling

In order to recognize a possible gamma-ray signal coming from Segue 1, it is necessary to know well enough the background in the *ON* region. As there is no possible way to measure this background directly, it is estimated from the simultaneously observed *OFF* region, where no gamma-ray sources are expected.

For the full likelihood analysis, it is needed to make an a priori assumption on the spectral shape of the emission from the *ON* region, for both its signal and background components (Eq. (4.6)). For the signal contribution, this is rather straightforward: once the dark matter model is chosen, the expected spectral shape (as a function of E') is convoluted with the response of the detector. As for the background, such approach cannot be taken, as the response function to the background emission is not known. A way around this problem is to estimate the shape of the background distribution as expected to be measured by the system, i.e. as a function of measured energy E.

The background in the *ON* region is approximated with the observations from the *OFF* region; therefore, it is the latter that has to be modeled for the purposes of the full likelihood calculations. In this work, it is done in the following way: four modeling zones are defined, of the same size and at the same angular distances from the camera center as the considered *OFF* region (Fig. 5.19). These zones are located adjacently to the *OFF*, two on each side, to ensure maximal agreement between them (in the sense of the camera homogeneity). Then, by means of the Kolmogorov-Smirnov (K-S) test, the energy distributions of events from each of the modeling zones (individually and

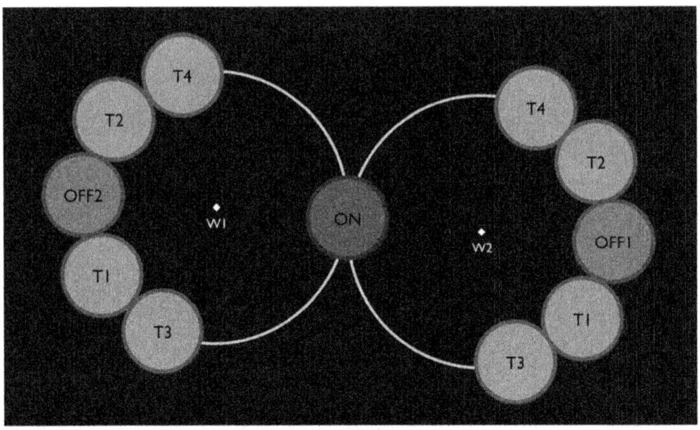

Fig. 5.19 Illustration of different zones used for background modeling. This particular scheme corresponds to sample C wobble configuration. See the main text for more details

combined), are compared to the energy distribution of events from the *OFF* region. The K-S statistics tests whether the events from two different sets come from the same parent distribution. If the calculated confidence level for their compatibility is close to 1 (with possible values between 0 and 1), the zone (or sum of several zones) in question is suitable for modeling the *OFF* region. The event distribution of most compatible zone (or combination of several) is then fitted with a continuous function, which is in turn used to model the background component in the full likelihood.

In this work, background is separately modeled for each wobble of each observational period: note that when telescopes are pointing to W1, data are gathered for the *OFF*2 region and the corresponding modeling zones (and vice versa: when pointing to W2, *OFF*1 and its corresponding modeling regions are recorded, Fig. 5.19). The K-S statistics is calculated by comparing the energy distribution of the *OFF* with the energy distributions of each of its four zones (T1, T2, T3 and T4), as well as the sums of two (T1 + T2), three (T1 + T2 + T3, T1 + T2 + T4) and all four (T1 + T2 + T3 + T4). The events from the zone(s) that turn out to be the most compatible (Table 5.7) are then used to model the background. In almost all the cases, it was empirically found that the distribution of those events can be nicely fitted by a smooth broken power law function (Fig. 5.20):

$$\frac{dN}{dE} = A_0 \left(\frac{E}{100\ \text{GeV}} \right)^{c_1} \left(1 + \left(\frac{E}{E_b} \right)^{\frac{c_1 - c_2}{\beta}} \right)^{-\beta}, \tag{5.4}$$

where A_0 is a arbitrary normalization constant, E_b is the break energy, β is the curvature of the transition between the low energy power law and high energy power law, whose respective indices are c_1 and c_2. Values of these parameters are estimated by a standard likelihood maximization, performed with the `TMinuit` class of ROOT [38]. The resulting best-fit parameter values, as well as the χ^2 estimates of such fits are listed in Table 5.7, for models of both *OFF* regions of the A, B1 and B2 observational periods.

For sample C, a different approach had to be applied. Namely, due to the dynamical sum cleaning, used for this data set (Sect. 5.2.3), the smooth broken power law no longer represents the optimal fit at the highest energies.[2] Therefore, in this case, the background is modeled directly from interpolation of the (measured) continuous, binned energy distribution of the events (Fig. 5.21) from the most compatible zone(s) (the K-S tests favor (T1 + T2 + T3 + T4) for *OFF*1 and (T1 + T2 + T3) for the modeling of *OFF*2).

Regarding the systematic uncertainties introduced in the results due to the choice of the background model, they are estimated in the following way: in accordance with the K-S statistics, the most and the least favored distributions are used for the modeling. These two different background models are then used in calculations of

[2] The dynamical sum cleaning is not the best solution for the Segue 1 sample C data, as it produces non-smooth response functions. However, by the time this work was finished, production of MC files with different cleanings was not yet possible.

Table 5.7 Details of the background modeling for the Segue 1 data samples A, B1 and B2

	Sample A		Sample B1		Sample B2	
	W1	W2	W1	W2	W1	W2
Zone	T1	T2	T1 + T2 + T3	T1 + T2 + T3	T1 + T2	T1 + T2
K-S c.l.	0.749	0.122	0.760	0.366	0.098	0.754
E_b [GeV]	73.57 ± 4.08	73.71 ± 3.25	71.97 ± 6.71	74.72 ± 2.88	72.94 ± 0.46	72.28 ± 0.53
c_1	3.87 ± 2.46	3.58 ± 2.42	5.16 ± 2.13	4.59 ± 2.22	6.00 ± 0.25	6.00 ± 0.37
c_2	-2.66 ± 0.04	-2.78 ± 0.04	-2.67 ± 0.04	-2.73 ± 0.04	-2.68 ± 0.02	-2.81 ± 0.03
β	0.39 ± 0.26	0.36 ± 0.21	0.46 ± 0.53	0.27 ± 0.23	0.77 ± 0.06	0.61 ± 0.08
χ^2	99.655	78.413	60.325	66.452	94.924	79.12
NDF	95	95	95	95	95	95
χ^2/NDF	1.049	0.825	0.635	0.699	0.999	0.832
Probability	0.491	0.891	0.998	0.988	0.624	0.939

Horizontal lines separate information regarding selection of the zone(s) used for modeling, sets of parameters obtained from the likelihood fit of the events from the zone(s), and values describing the validity of the fit

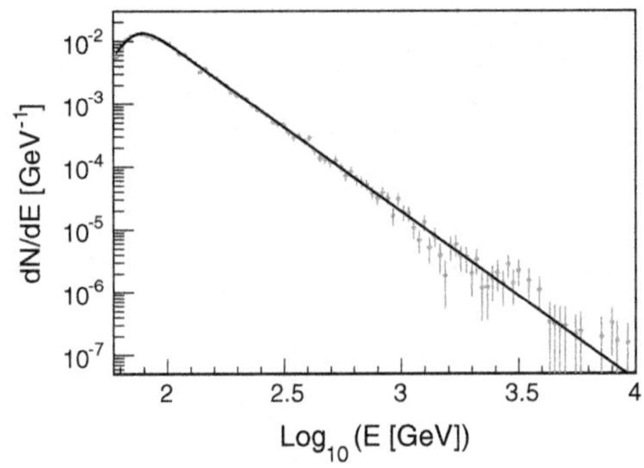

Fig. 5.20 Example of background modeling with the smooth broken power law (applied for samples A, B1 and B2). The *green* points represent events from the zone(s) most compatible with the *OFF* region, and the *black line* is the fit to their distribution. This particular case represents the background modeling of *OFF2* region for the sample B2. (Color in online)

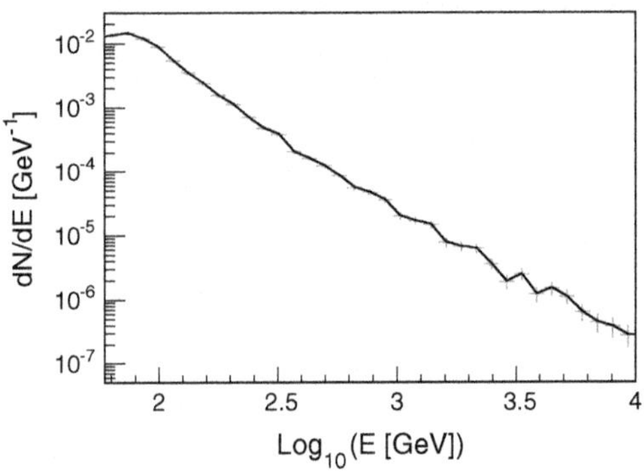

Fig. 5.21 Example of background modeling for the Segue 1 sample C data. The *green* points represent events from the zone(s) most compatible with the *OFF* region, and the *black line*, used for the modeling, is obtained from the extrapolation of these points. This particular case refers to the background modeling of *OFF1* region. See the main text for more details. (Color in online)

the $\langle\sigma_{\rm ann}v\rangle$ and τ_χ limits for a given dark matter scenario. The maximal discrepancies between the obtained results are taken as an estimate of the systematic uncertainties: for $m_\chi < 300$ GeV, the systematics are (as expected) relatively high—up to 47 %. At 1 TeV, the uncertainties are ~5 %, slightly increasing again for the highest energies, up to ~15 % for $m_\chi = 10$ TeV.

5.3.5 Signal

Lastly, the spectral function of the expected gamma-ray emission from the source has to be defined. For the work presented here, this means selecting a certain model of dark matter annihilation or decay for which investigation with MAGIC makes sense. Once the model is chosen, the spectral shape of the signal is fixed by the dN/dE' term of the particle physics flux contribution Φ^{PP} (Eqs. (2.2 and 2.6)). This spectral shape, however, is defined as a function of true energy E'; to transform it to the function of measured energy E, it has to be convoluted with the response of the telescopes (Eq. (4.6)).

Some of the mechanisms for the production of gamma-rays through the annihilation or decay of dark matter particles have already been presented in Sect. 2.3.2.3. This section lists the particular models that will be considered in this work, and presents some of the more technical aspects regarding their analysis. In order to make the results as model-independent as possible, in all the cases, the branching ratio is set to Br = 100 %, and the dark matter particle mass m_χ takes values from 100 GeV−10 TeV (200 GeV−20 TeV for the decay scenarios).

Secondary Photons, produced from annihilation or decay of dark matter into SM particles (with FSR included), have continuous and rather featureless spectral distributions, with a cutoff at the kinematical limit $E_\gamma = m_\chi$ (Sect. 2.3.2.3). Thus, after the convolution with the response function, the principal change in their spectra is the alteration of the slope, softening of the cutoff and a slight extension to energies beyond m_χ (Fig. 5.22). The following final state products are considered:

- quarks: $b\bar{b}$ and $t\bar{t}$;
- leptons: $\mu^+\mu^-$ and $\tau^+\tau^-$; and
- gauge bosons: W^+W^- and ZZ.

Their modeling is done from the fit functions presented in [39].

Gamma-Ray Lines, resulting from direct annihilation or decay of dark matter particles into photon(s), are expected to be the 'smoking guns' of indirect searches (Sect. 2.3.2.3). The monochromatic line is in practice modeled as a very narrow gaussian, of relative width $\sim 10^{-3}$ (due to the Doppler shift) and centered at m_χ. However, due to the finite energy resolution of the instruments, if detected, line would be seen as widened and smoothed (Fig. 5.23). In this work, line is expected from the following annihilation processes:

- $\chi\chi \to \gamma\gamma$, with $E_\gamma = m_\chi$ and
- $\chi\chi \to Z\gamma$, with $E_\gamma = m_\chi\left(1 - m_Z^2/4m_\chi^2\right)$.

For the case of the decay into $\gamma\gamma$ or $Z\gamma$, the line is centered at $E_\gamma = m_\chi/2$ or $E_\gamma = m_\chi\left(1 - m_Z^2/4m_\chi^2\right)/2$, respectively.

Virtual Internal Bremsstrahlung photons have spectral distribution characterized by the prominent peak at energies close to m_χ, and by the sharp cutoff at $E_\gamma = m_\chi$ (Figs. 2.8, 5.24). As these photons are emitted in dark matter annihilation into charged SM particles, the total spectrum is composed of continuous distributions of secondary photons and FSR, plus the VIB bump; therefore, in this work, limits are set on the total 3-body annihilation cross section. Because of the finite response function of the telescopes, the shape of the VIB part of the spectrum is widened by the energy resolution, the cutoff is softened and spectrum extended above m_χ (Fig. 5.24).

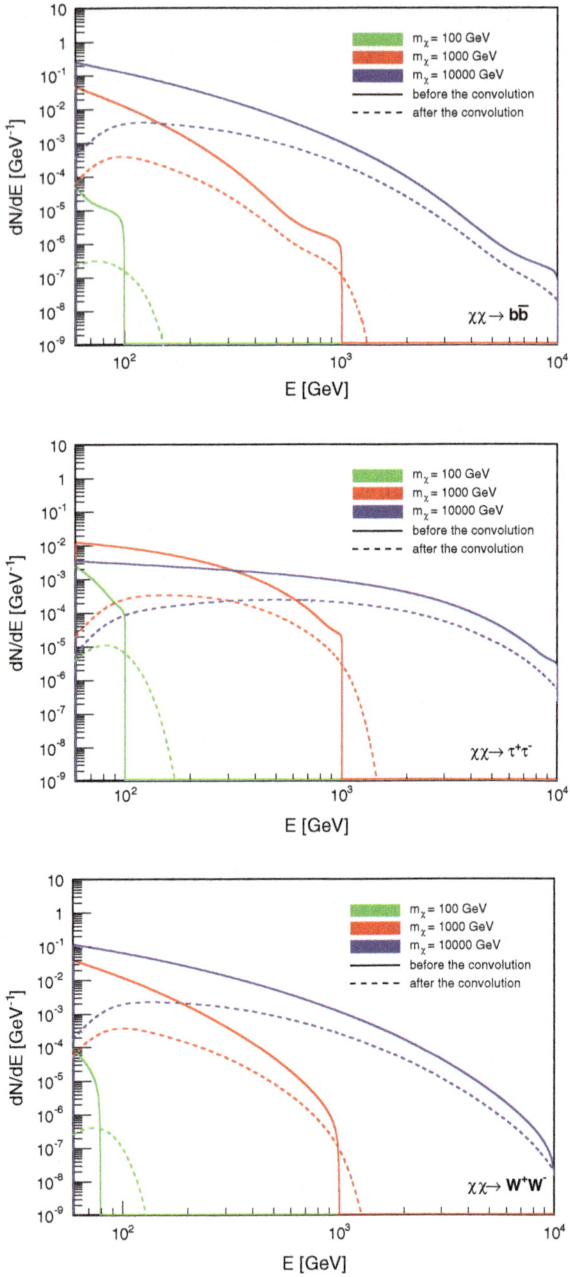

Fig. 5.22 Gamma-ray spectra from dark matter annihilation into $b\bar{b}$ (*top*), $\tau^+\tau^-$ (*middle*) and W^+W^- (*bottom*) channels (*full lines*) for m_χ = 100, 1000 and 10000 GeV (*green*, *red* and *violet lines*, respectively). The *dashed lines* represent the shapes of the spectra convoluted with the response function of the MAGIC Telescopes

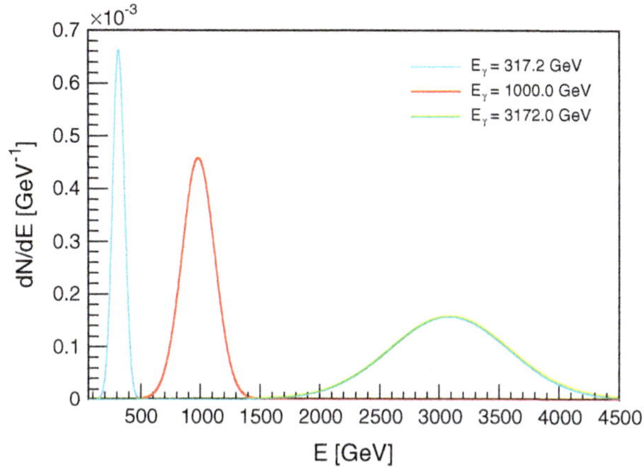

Fig. 5.23 Differential energy spectrum of gamma-ray lines after the convolution with the MAGIC response function. The shown example assumes direct dark matter annihilation into two photons, with m_χ taking values of 317.2, 1000 and 3172 GeV (*blue*, *red* and *green lines*, respectively)

Fig. 5.24 Continuous gamma-ray spectrum with VIB contribution from dark matter annihilation into $\mu^+\mu^-(\gamma)$ (*top*) and $\tau^+\tau^-(\gamma)$ (*bottom*), with m_χ taking values of 317.2, 1000 and 3172 GeV (*blue*, *red* and *green lines*, respectively). Considered are the cases of the mass-splitting parameter μ being 1.05 (*lighter colors*) or 2.00 (*darker colors*). *Full lines* represent the spectra before, and *dashed lines* after the convolution with the response function of MAGIC

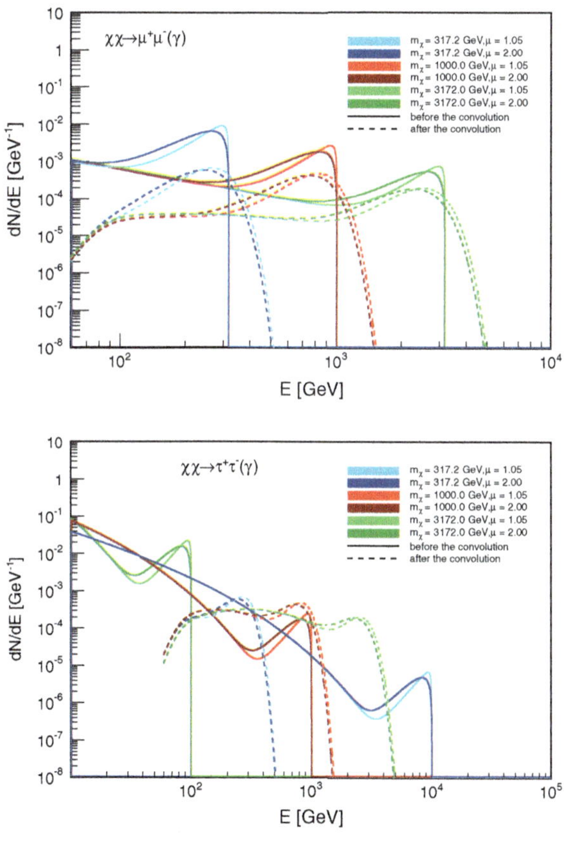

In this work, the annihilation into leptonic channels is considered:

- $\chi\chi \rightarrow \mu^+\mu^-(\gamma)$, and
- $\chi\chi \rightarrow \tau^+\tau^-(\gamma)$.

How pronounced is the VIB bump depends on the mass-splitting parameter μ ($\equiv m_\eta^2/m_\chi^2$), where η is the t-channel mediator (Sect. 2.3.2.3). For small μ values the gamma-ray spectrum at high energies is completely dominated by VIB photons, while the secondary photons and FSR are relevant at lower energies (or for large values of μ). Here are considered the VIB contributions for $1 < \mu \leq 2$, as virtual photons are the most important in the nearly degenerate cases; furthermore, for this range, the spectral shape of the VIB signal is practically independent of μ [40].

Gamma-ray Boxes are other class of possibly pronounced spectral features from dark matter annihilation or decay (Sect. 2.3.2.3). Assuming that the dark matter particles annihilate (decay) into photons via intermediate scalars ϕ, the resulting gamma-ray emission is isotropic, and the spectral shape is constant between the endpoints—like a box (Fig. 5.25). Center and width of the box are entirely fixed by the masses

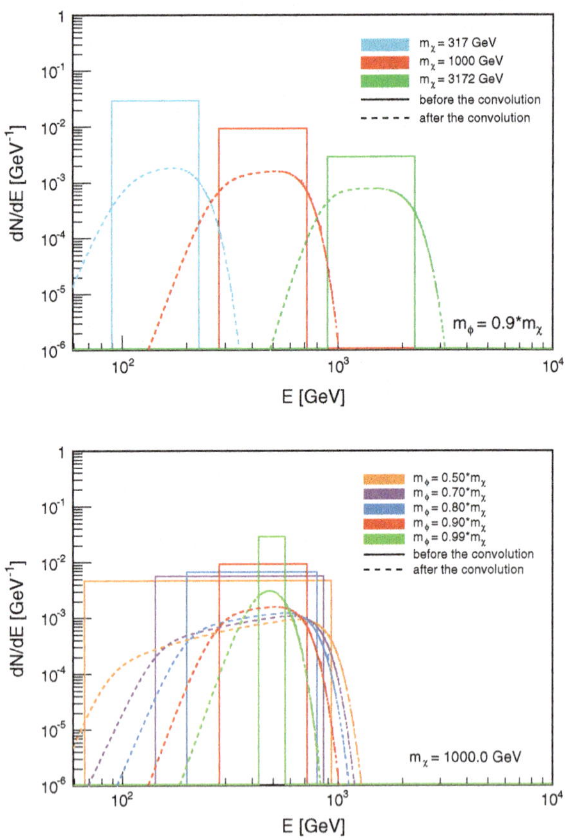

Fig. 5.25 Box-like gamma-ray spectra from dark matter annihilation, when m_χ takes values of 317.2, 1000 and 3172 GeV (*blue*, *red* and *violet lines*, respectively) and $m_\phi = 0.90 * m_\chi$ (*top*), and when m_χ = 1000 GeV while the mass of the scalar ϕ varies as m_ϕ/m_χ = 0.5, 0.7, 0.8, 0.9 and 0.99 (*orange*, *violet*, *blue*, *red* and *green lines*, respectively; *bottom*). *Full lines* represent the spectra before, and *dashed lines* after the convolution with the response function of MAGIC

of the dark matter and scalar particles, m_χ and m_ϕ, respectively: the box is centered at $E_c = m_\chi/2$, and its width is given as $\sqrt{(m_\chi^2 - m_\phi^2)}$. For $m_\phi \approx m_\chi$, almost all of the dark matter particle energy is transferred to the photons, and the resulting spectral shape is intense and similar to the monochromatic line. On the other hand, for $m_\phi \ll m_\chi$, the box becomes wide and dim in amplitude; still, it extends to high energies and thus is not negligible as a contribution to the signal spectrum.

Convolution of the box-like spectral shape with the response function of the telescopes smoothens the edges and broadens the signal distribution; in addition, because the energy resolution depends on E', the convoluted box is not symmetric around E_c (Fig. 5.25).

Here is considered the case of scalar ϕ decaying into two photons ($\phi \rightarrow \gamma\gamma$), meaning that annihilation of two dark matter particles results in production of total of 4 photons. As for the dark matter decay scenarios, they are not presented in this work, given that the transformation of $\langle\sigma_{\rm ann}v\rangle$ upper limits to the τ_χ lower limits is trivial, by making the replacements $\langle\sigma_{\rm ann}v\rangle/8\pi m_\chi^2 \rightarrow 1/m_\chi\tau_\chi$ and $m_\chi \rightarrow m_\chi/2$ (the same conversion is applied for the cases of decay into final state SM particles (Sect. 5.4.1.2) and the monochromatic line (Sect. 5.4.2.2)).

5.3.6 The Astrophysical Factor *J* for Segue 1

In order to estimate the expected gamma-ray flux from photons of dark matter origin, it is needed to know the value of the astrophysical term J, calculated from Eqs. (2.4) and (2.7), for the annihilation and decay scenarios, respectively.

In this work, the dark matter density distribution is modeled assuming the Einasto profile (Sect. 2.3.2.1, [41]), with the scale radius of $r_s = 0.15$ kpc, scale density of $\rho_s = 1.1 \times 10^8\ M_\odot\ \text{kpc}^{-3}$ and slope $\alpha = 0.30$ [25]. The solid angle over which ρ^2 (for the annihilation) or ρ (for the decay) is integrated, $\Delta\Omega$, is defined by the θ^2-cut optimized for this analysis: $\Delta\Omega = 2\pi\left(1 - \cos(\theta)\right)$. Figure 5.26 shows the value of J as a function of the applied cut in θ: the greater the area treated as the source region, the greater the J value.

For the given analysis cuts, and for models where dark matter particles annihilate, the J factor is $J = 1.05 \times 10^{19}\,\text{GeV}^2\ \text{cm}^{-5}$ sr. As for the decay scenarios, the corresponding value is $J = 2.92 \times 10^{17}\,\text{GeV cm}^{-2}$ sr. The systematic uncertainties on the J value, resulting from the fit of the Segue 1 dark matter distribution to an Einasto profile are about an order of magnitude at the 1σ level [42]. These uncertainties affect the $\langle\sigma_{\rm ann}v\rangle$ or τ_χ limits linearly.

5.3.7 Analysis Technicalities

For each of the two wobble positions of each of the four defined observational periods (Sect. 5.2), a PDF for the full likelihood analysis is constructed using the

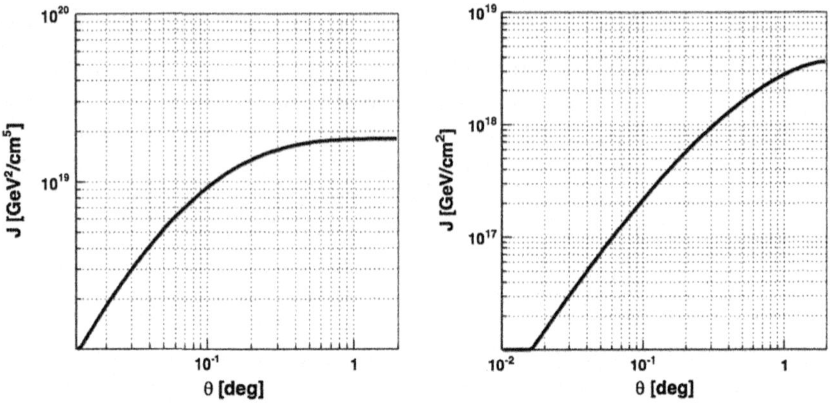

Fig. 5.26 Astrophysical factor J for Segue 1, assuming Einasto density profile and dark matter annihilation (*left*) and decay (*right*) scenarios, as a function of θ analysis cut

corresponding background model (Sect. 5.3.4) plus the signal spectral function (Sect. 5.3.5) convoluted with the appropriate response of the telescopes (Sect. 5.3.3). Each PDF is then used to calculate the likelihood for the given data set of the Segue 1 observations, previously reduced by the application of optimized cuts (Sect. 5.3.1).

The full likelihood calculations are performed for the 95 % c.l. and one-sided confidence intervals using the `TMinuit` class of ROOT [43]. The gamma-ray signal intensity, A, is included as a free parameter, and its limit value is obtained as a sum of its minimum $A_{\rm min}$ (that corresponds to the maximum likelihood value) plus the width of the confidence interval $A_{\rm CI}$. A limitation is set, so that A can only assume values in the physical region, i.e. it can not be negative. The results obtained this way are conservative, since negative statistical (or systematic) fluctuations cannot produce artificially constraining limits. Additionally, the number of estimated background events b (related to the number of estimated gamma-ray events g through Eq. (4.9)) and the background normalization τ are treated as the nuisance parameters in the likelihood maximization.

These calculations are made for each of the two wobble positions of each of the four observational samples, but also for the entire Segue 1 data sample. The latter is done by combining the above mentioned 8 individual likelihoods into an overall one, following Eq. (4.15).

Lastly, the conversion of A value into $\langle\sigma_{\rm ann}v\rangle$ or τ_χ is rather straightforward: from Eqs. (4.3–4.6), A can be expressed as a function of $d\Phi/dE$, which is then replaced by the corresponding form of the expected gamma-ray flux (Eq. (2.1)).

5.4 Results

Finally, this section brings the results of this work—interpretation, in the context of indirect dark matter searches, of 157.9 h of good-quality data from the Segue 1 observations with MAGIC, analyzed with the full likelihood approach.

Results are presented in the following way: for each of the dark matter models listed in Sect. 5.3.5, a limit is set (upper limit on $\langle\sigma_{\rm ann}v\rangle$ or lower limit on τ_χ) by combining the likelihood analyses of the whole data sample. This joint constraint is then compared to the expectations (from fast MC simulations) for the null hypothesis (no signal), as well as for signals of 1σ and 2σ significances. A deviation from the null hypothesis could suggest a hint of a signal, but only if it is sufficiently strong, with the look-elsewhere or trial factor taken into account. In this work all limits are bellow $\sim 2\sigma$; therefore, no statistically significant signal detection can be claimed.

Additionally, when applicable, the bounds from this work are compared to the currently most constraining results from other gamma-ray observatories. All of the results are derived without the assumptions of some additional boosts, either from the presence of the substructures (Sect. 2.3.2.1) or from quantum effects (Sect. 2.3.2.2).

As for the uncertainties affecting the obtained limits, they can be grouped as:

- systematic uncertainties from background modeling; they are $\sim$50 % below 300 GeV, $\sim$5 % at 1 TeV and up to 15 % above 5 TeV (Sect. 5.3.4);
- uncertainty on density profile parameters, translated to about one order of magnitude error at the 1σ level on the J factor value (Sect. 5.3.6).

5.4.1 Secondary Photons from Final State SM Particles

5.4.1.1 Annihilation

Upper limits on $\langle\sigma_{\rm ann}v\rangle$ are computed for each of the six considered annihilation channels ($b\bar{b}$, $t\bar{t}$, $\mu^+\mu^-$, $\tau^+\tau^-$, W^+W^- and ZZ). They are presented separately and in comparison with the most stringent bounds from Fermi-LAT, H.E.S.S. and VERITAS, described in the following paragraphs.

The Fermi-LAT results are from the already mentioned joint analysis of 2 years of observations of 10 dSph galaxies (Figs. 2.12 and 5.3, [22]). The obtained limits assume dark matter particle annihilation, with Br = 100 %, into $b\bar{b}$, $\mu^+\mu^-$, $\tau^+\tau^-$ and W^+W^- final states, for m_χ in the 5 GeV–1 TeV range (100 GeV–1 TeV for W^+W^-). For $b\bar{b}$ and $\tau^+\tau^-$ channels, and WIMPs lighter than 25 GeV and 40 GeV, respectively, the Fermi-LAT constraints on $\langle\sigma_{\rm ann}v\rangle$ are actually stronger than the thermal rate of $\langle\sigma_{\rm ann}v\rangle \sim 3\times10^{-26}$ cm^3 s^{-1} (Fig. 2.12).

For higher m_χ values, however, the Fermi-LAT limits become significantly weaker, and above 700 GeV, currently the strongest bounds are derived from 112 h of H.E.S.S. observations of the Galactic Halo [44]. Assuming a generic model of dark matter annihilation into quarks, for $m_\chi = 1$ TeV, $\langle\sigma_{\rm ann}v\rangle$ is only a factor of $\sim$10 away from the thermal $\langle\sigma_{\text{ann}}v\rangle$ (Fig. 2.10). Still, when comparing limits from different targets, one should always keep in mind that the underlying astrophysical uncertainties in calculation of the factor J may be quite different: for instance, predictions for integrated signal fluxes are much more robust for dSphs than for the Galactic Center.

As for the limits from observations of dSphs by the IACTs, the most constraining ones were published by the VERITAS Collaboration in 2012 [27]. From ~48 h of Segue 1 data, $\langle\sigma_{\text{ann}}v\rangle$ exclusion curves were derived for $b\bar{b}$, $\mu^+\mu^-$, $\tau^+\tau^-$ and W^+W^- channels, assuming m_χ in the 100 GeV – 11 TeV range. However, the validity of these results is under doubt: the cross-check of those bounds, done by the Author of this work and her collaborators [45], indicate some inconsistencies in the VERITAS publication [27], most likely caused by an overestimation of the performance of the VERITAS telescopes. Namely, for even the most conservative assumptions regarding the response function of VERITAS, the numbers provided in [27] imply a sensitivity of these telescopes of 0.32 % of the Crab Nebula flux above 165 GeV, more than a factor ~2 better than officially claimed by the VERITAS Collaboration [46]. As a consequence, all of their $\langle\sigma_{\text{ann}}v\rangle$ limits are probably overestimated. Therefore, although the VERITAS results from Segue 1 will be shown on the following plots, together with the constraints from Fermi-LAT, H.E.S.S. and from this work, they will not be discussed and should be regarded by the Reader with serious reserves.

Lastly, constraints from ~30 h of the 2009 observations of Segue 1 with MAGIC in the single telescope mode (MAGIC-II was still under commissioning at this period) are considered [25]. Limits were derived for dark matter particle annihilation into $b\bar{b}$, $\tau^+\tau^-$ and $\mu^+\mu^-$ channels, with m_χ between 200 GeV and 1350 GeV. These results are included in this section in order to illustrate the advance that has been achieved in the dark matter searches with MAGIC since 4 years ago, by means of deeper observations, more sensitive instruments and dedicated analysis approach.

- $\chi\chi \to b\bar{b}$

As already mentioned in Sect. 5.3.7, the $\langle\sigma_{\text{ann}}v\rangle$ exclusion curve is obtained by combining the likelihood of each of the wobble positions of each of the four observational periods. For illustration purposes, these individual limits are shown on Fig. 5.27. As it can be seen, the weakest constraints come from the statistically 'poor' B1 samples, while the fully upgraded system (sample C) yields the strongest bounds.[3] Also shown is the combined upper limit on $\langle\sigma_{\text{ann}}v\rangle$—it is more stringent, by at least a factor ~2, than the most constraining individual bound at any energy.

Figure 5.28 shows the combined upper limit on $\langle\sigma_{\text{ann}}v\rangle$ calculated from the Segue 1 data (full line), together with the null-hypothesis expectations (dashed line), and expectations for 1σ (shaded gray area) and 2σ (shaded light blue area) significant signal. The limit from this work is consistent with a no-detection scenario. The fact that it agrees so well with the null signal is due to the fact that the free parameter is bounded within the physical region during the fit (Sect. 5.3.7).

[3] Above 2 TeV, these limits are somewhat deteriorated, as the dynamical sum image cleaning is no longer optimal for this analysis.

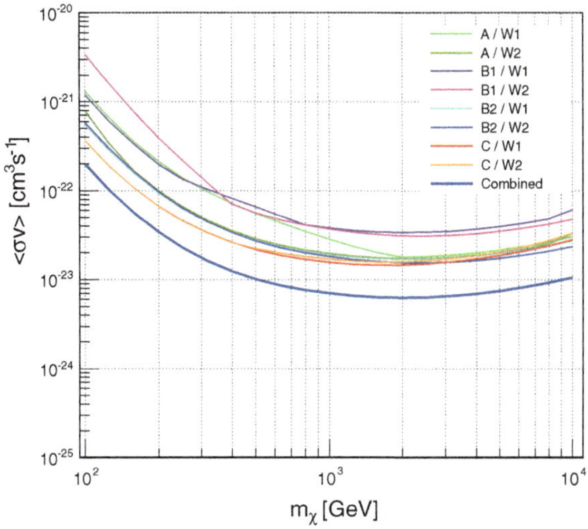

Fig. 5.27 Upper limits on $\langle\sigma_{\text{ann}} v\rangle$ for $b\bar{b}$ annihilation channel, from individual wobble positions and different Segue 1 observational periods. Also shown is the limit from the combined likelihood analysis

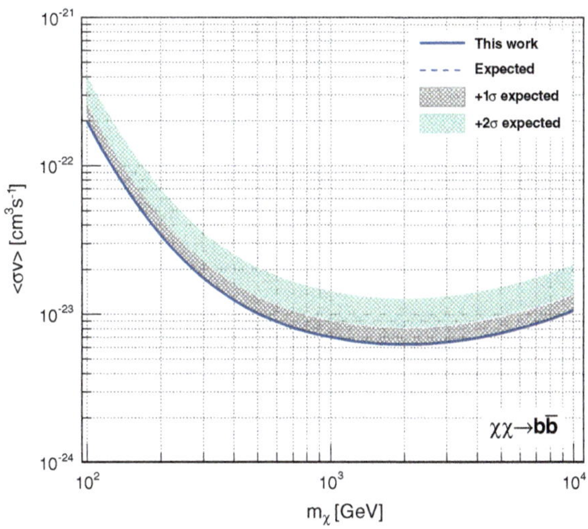

Fig. 5.28 Upper limits on $\langle\sigma_{\text{ann}} v\rangle$ for $b\bar{b}$ channel as a function of m_χ, from the Segue 1 observations with MAGIC (*full line*) and as expected for the case of no signal (*dashed line*), or for the signal with 1σ or 2σ significance (*gray* and *light blue shaded areas*, respectively)

The $\langle\sigma_{\text{ann}} v\rangle$ limit from different experiments are compared in Fig. 5.29: the result from this work is the strongest limit from dSph observations by the IACTs, reaching a minimum of $\langle\sigma_{\text{ann}} v\rangle = 6.2 \times 10^{-24}$ cm^3 s^{-1} for $m_\chi = 2$ TeV. It complements the Fermi-LAT constraints above 1 TeV, and it is more than an order of magnitude better than the previous MAGIC limit from mono observations.

The most stringent bound for $m_\chi > 700$ GeV is still that from the H.E.S.S observations of the Galactic Halo (for a channel of similar spectral shape as $\chi\chi \rightarrow b\bar{b}$). It is, however, derived for a different target with much higher J factor than Segue 1, but also potentially affected by very different systematic uncertainties.

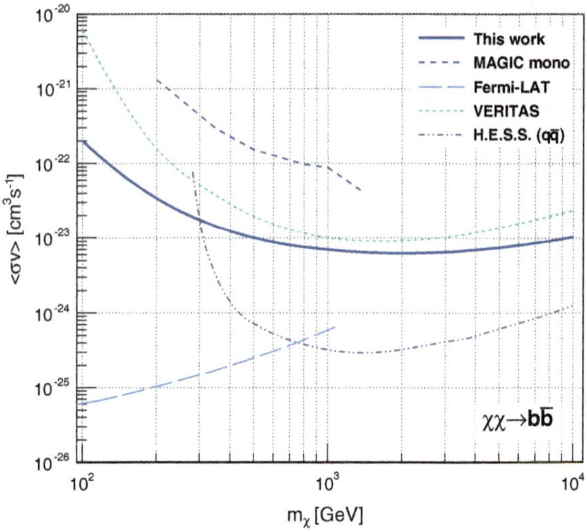

Fig. 5.29 Upper limits on $\langle\sigma_{\rm ann}v\rangle$ for $b\bar{b}$ channel as a function of m_χ, from this work (*full line*), compared with the exclusion curves from Fermi-LAT [22], MAGIC-I [25] VERITAS [27] and H.E.S.S. [44] (*long-dashed, dashed, dotted* and *dashed-dotted lines*, respectively)

- $\chi\chi \rightarrow t\bar{t}$

This annihilation channel becomes an option for $m_\chi > m_{\rm t} = 173.5 \pm 0.6$ GeV. In this work, the upper limit on $\langle\sigma_{\rm ann}v\rangle$ for $t\bar{t}$ final state is calculated for $m_\chi > 200$ GeV. Figure 5.30 compares the result from observations with the expectations; again, there is no hint of a signal. No experimental results from other gamma-ray observatories have been found for this channel. The strongest limit is calculated for $m_\chi = 2.5$ TeV, and it is of order of $\langle\sigma_{\rm ann}v\rangle = 8.7 \times 10^{-24}$ cm^3 s^{-1}.

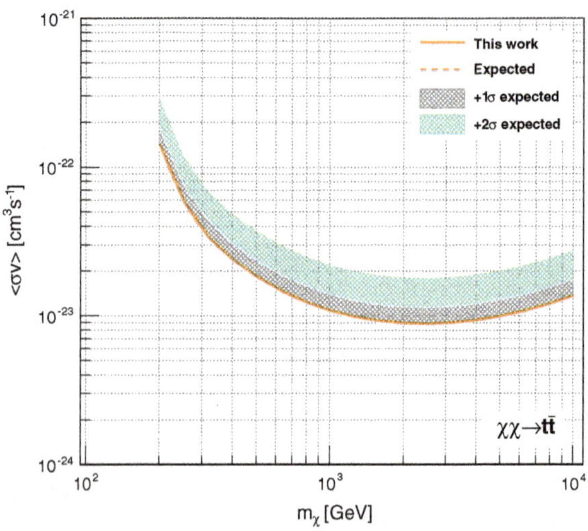

Fig. 5.30 Upper limits on $\langle\sigma_{\rm ann}v\rangle$ for $t\bar{t}$ channel as a function of m_χ, from the Segue 1 observations with MAGIC (*full line*) and as expected for the case of no signal (*dashed line*), or for the signal with 1σ or 2σ significance (*gray* and *light blue shaded areas*, respectively)

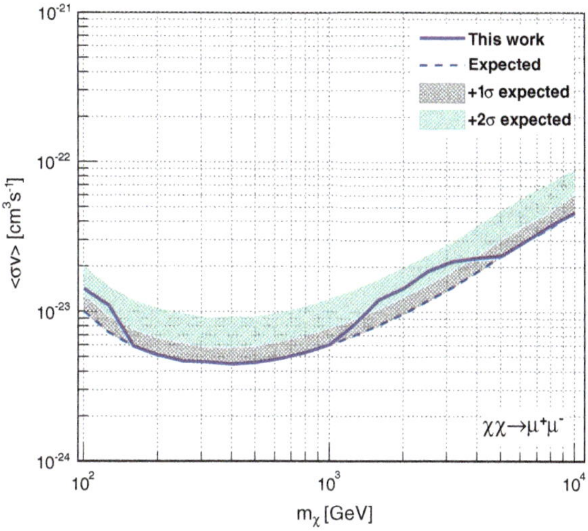

Fig. 5.31 Upper limits on $\langle\sigma_{\rm ann}v\rangle$ for $\mu^+\mu^-$ channel as a function of m_χ, from the Segue 1 observations with MAGIC (*full line*) and as expected for the case of no signal (*dashed line*), or for the signal with 1σ or 2σ significance (*gray* and *light blue shaded areas*, respectively)

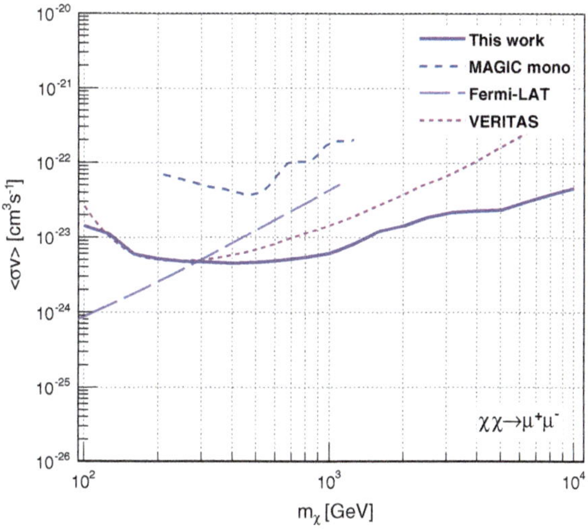

Fig. 5.32 Upper limits on $\langle\sigma_{\rm ann}v\rangle$ for $\mu^+\mu^-$ channel as a function of m_χ, from this work (*full line*), compared with the exclusion curves from Fermi-LAT [22], MAGIC-I [25] and VERITAS [27] (*long-dashed*, *dashed* and *dotted lines*, respectively)

- $\chi\chi \to \mu^+\mu^-$

Gamma-ray photons from the $\mu^+\mu^-$ channel are emitted from the FSR process only. Figure 5.31 shows $\langle\sigma_{\rm ann}v\rangle$ upper limits, from observations and from expectations. There are some positive statistical fluctuations, but no signal can be claimed.

Figure 5.32 compares the result from this work with those from Fermi-LAT: for $m_\chi > 300$ GeV, MAGIC limit becomes the most constraining, with the strongest value of $\langle\sigma_{\rm ann}v\rangle = 4.4\times10^{-24}$ cm^3 s^{-1} corresponding to $m_\chi = 400$ GeV. This limits also represents an improvement with respect to the MAGIC mono result by one order of magnitude.

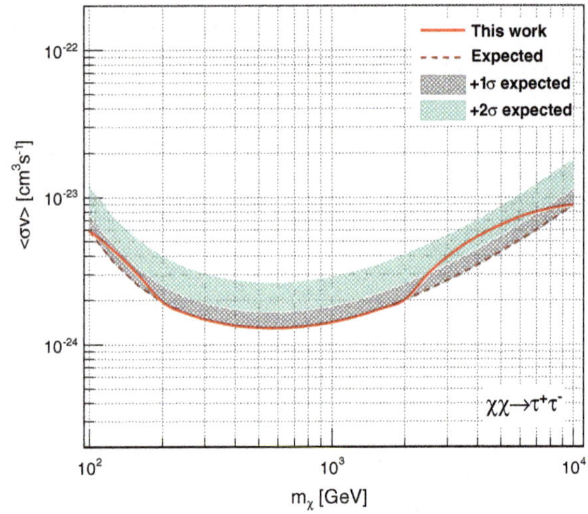

Fig. 5.33 Upper limits on $\langle\sigma_{\mathrm{ann}}v\rangle$ for $\tau^+\tau^-$ channel as a function of m_χ, from the Segue 1 observations with MAGIC (*full line*) and as expected for the case of no signal (*dashed line*), or for the signal with 1σ or 2σ significance (*gray* and *light blue shaded areas*, respectively)

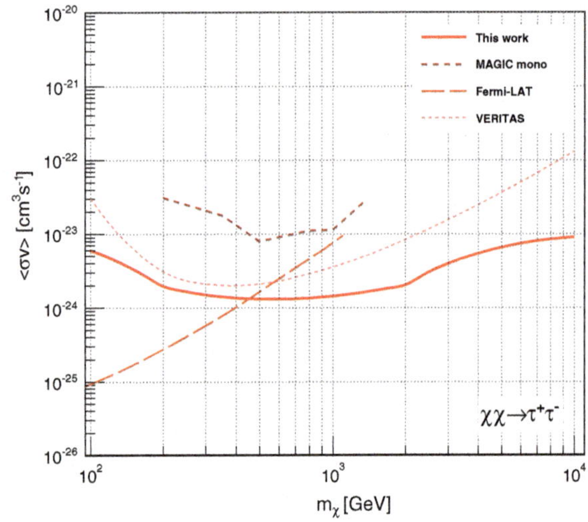

Fig. 5.34 Upper limits on $\langle\sigma_{\mathrm{ann}}v\rangle$ for $\tau^+\tau^-$ channel as a function of m_χ, from this work (*full line*), compared with the exclusion *curves* from Fermi-LAT [22], MAGIC-I [25] and VERITAS [27] (*long-dashed*, *dashed* and *dotted lines*, respectively)

- $\chi\chi \to \tau^+\tau^-$

The upper limits on $\langle\sigma_{\mathrm{ann}}v\rangle$ for this channel show some statistical fluctuations at high m_χ values, but no hint of a signal (Fig. 5.33). The exclusion curve from this work becomes more stringent than the one from Fermi-LAT for $m_\chi > 450$ GeV (Fig. 5.34). It is at the same time the most constraining result from dSphs observations with IACTs till date, and the strongest limit from this work for annihilation to secondary photons, with the lowest $\langle\sigma_{\mathrm{ann}}v\rangle$ value of 1.2×10^{-24} cm^3 s^{-1} for $m_\chi = 550$ GeV—a factor $\sim$40 away from the thermal rate of 3×10^{-26} cm^3 s^{-1}, and almost an order of magnitude below the MAGIC mono bound.

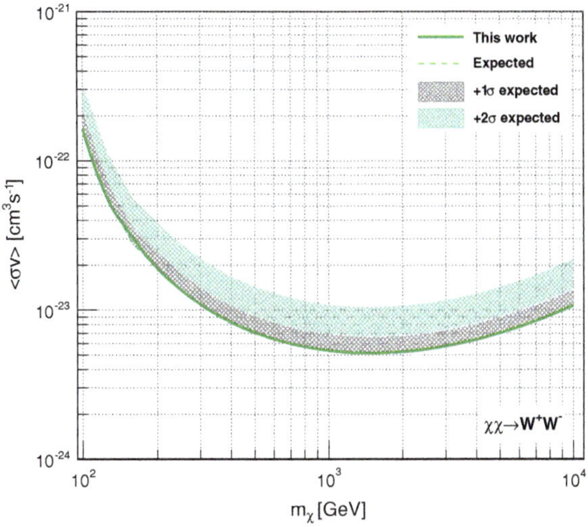

Fig. 5.35 Upper limits on $\langle\sigma_{\rm ann}v\rangle$ for W^+W^- channel as a function of m_χ, from the Segue 1 observations with MAGIC (*full line*) and as expected for the case of no signal (*dashed line*), or for the signal with 1σ or 2σ significance (*gray* and *light blue shaded areas*, respectively)

- $\chi\chi \rightarrow W^+W^-$

The upper limit on $\langle\sigma_{\rm ann}v\rangle$ for W^+W^- final state is shown in comparison with different expectations in Fig. 5.35, and with the results from different experiments in Fig. 5.36. It is the most constraining result from observations of dSphs above 1 TeV, with the most stringent bound of $\langle\sigma_{\rm ann}v\rangle = 5.1 \times 10^{-24}$ cm^3 s^{-1} corresponding to $m_\chi = 1.5$ TeV.

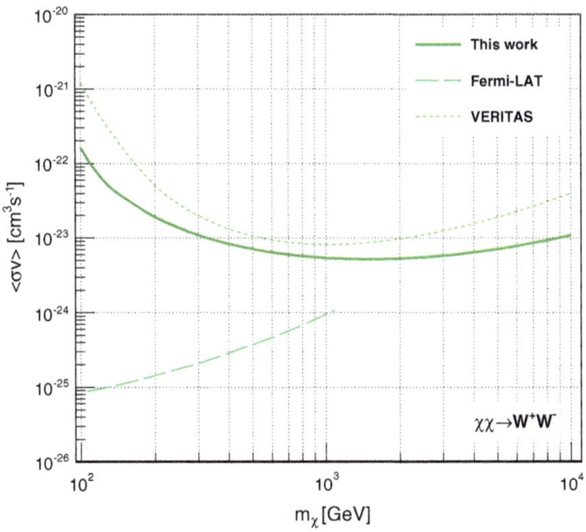

Fig. 5.36 Upper limits on $\langle\sigma_{\rm ann}v\rangle$ for W^+W^- channel, from this work (*full line*), compared with the exclusion *curves* from Fermi-LAT [22] and VERITAS [27] (*long-dashed* and *dotted lines*, respectively)

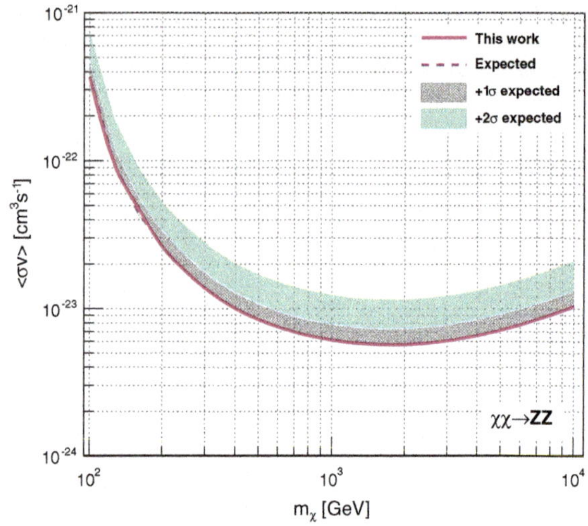

Fig. 5.37 Upper limits on $\langle\sigma_{\text{ann}} v\rangle$ for ZZ channel as a function of m_χ, from the Segue 1 observations with MAGIC (*full line*) and as expected for the case of no signal (*dashed line*), or for the signal with 1σ or 2σ significance (*gray* and *light blue shaded areas*, respectively)

- $\chi\chi \to ZZ$

Although intrinsically different, channels ZZ and W^+W^- have very similar spectral shapes, thus it is not surprising that their limits are similar as well. Figure 5.37 shows the upper limit for $\langle\sigma_{\text{ann}} v\rangle$ together with the expectations for ZZ channel. The most constraining result corresponds to $m_\chi = 1.7\,\text{TeV}$, and is of order $\langle\sigma_{\text{ann}} v\rangle \sim 5.6 \times 10^{-24}\,\text{cm}^3\,\text{s}^{-1}$. No constraints from other gamma-ray observatories have been found for this channel.

Lastly, for a more comprehensive overview, all of the $\langle\sigma_{\text{ann}} v\rangle$ upper limits obtained from this work for the considered final state channels are plotted in Fig. 5.38. These are the strongest constraints from dSphs observations with IACTs; Depending on the channel, they are a factor ~40–300 away from the $\langle\sigma_{\text{ann}} v\rangle$ canonical value.

5.4.1.2 Decay

Over the last couple of years, a lot of attention has been given to the decaying dark matter as a possible explanation of the flux excesses of high-energy positrons and electrons measured by AMS-02, PAMELA, Fermi-LAT and H.E.S.S (see, e.g. [47, 48]). The needed dark matter particle lifetime in such case, $\tau_\chi > 10^{26}$, is much longer than the age of the Universe so that the slow decay does not significantly reduce the overall dark matter abundance and, therefore, there is no tension with the astrophysical and cosmological observations.

Although dark matter decay has been abandoned as the cause of the observed flux excesses in favor of some better justified, astrophysical explanations, this scenario is an interesting subject by itself: if the dark matter particle, for example, is a gravitino, and small R-parity violation is allowed, the appropriate relic density through the

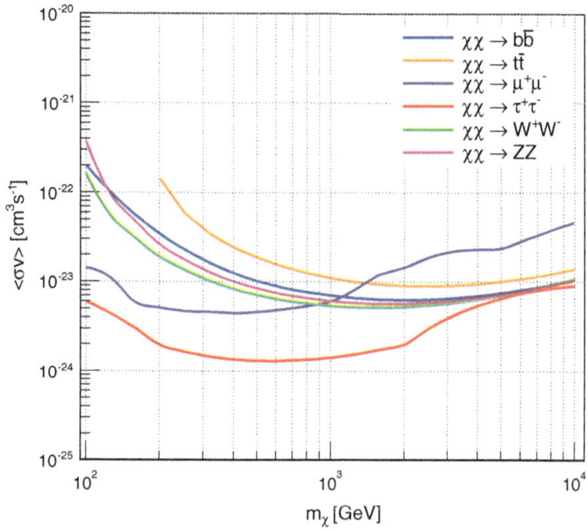

Fig. 5.38 Upper limits on $\langle\sigma_{\rm ann} v\rangle$, for different final state channels, from Segue 1 observations with MAGIC

thermal production is achieved, naturally leading to a cosmological history consistent with thermal leptogenesis and primordial nucleosynthesis [49].

The currently strongest constraints on τ_χ from gamma-ray observatories are derived from the Fermi-LAT isotropic gamma-ray data: the 2-year long measurements extending from ~200 MeV up to 580 GeV [50] exclude decaying dark matter with lifetime shorter than 10^{26} s (depending on m_χ and the channel, [51]). As for the IACTs, recent observations of the Fornax galaxy cluster by H.E.S.S. (~15 h of data, $E > 100$ GeV, [52]) have been used to exclude dark matter models for which $\tau_\chi < (10^{25}-10^{26})$ s [51].

As far as the suitable targets for decaying dark matter searches go, dSphs are a suboptimal choice compared to the galaxy clusters or to the whole sky (Sect. 2.3.3). Still, it is interesting to see how limits for the same channel of dark matter decay compare from observations of different objects and by different observatories.

In this section, models of the 2-body dark matter decay into six final states, typical of scalar dark matter, are considered: $b\bar{b}$, $t\bar{t}$, $\mu^+\mu^-$, $\tau^+\tau^-$, W^+W^- and ZZ (Sect. 5.3.5), plus the FSR when possible. The 3-body decays are left out as model-dependent. For fermionic dark matter, decay channels such as $\chi \to l^\pm W^\mp$ and $Z\nu$ are possible, but these (in first approximation) can be analyzed as combination of the corresponding 2-body scalar dark matter decay channels.

Lower limits on τ_χ are presented for each channel separately, in comparison with the corresponding bounds from the above mentioned Fermi-LAT and H.E.S.S. data (when applicable). For the $\mu^+\mu^-$ and $\tau^+\tau^-$ channels, also shown are the best fits to the Fermi and PAMELA measurements of the positron fraction [47].

- $\chi \to b\bar{b}$

Figure 5.39 shows the lower limit on τ_χ from Segue 1 observations with MAGIC. The strongest bound corresponds to the most massive (considered) dark matter

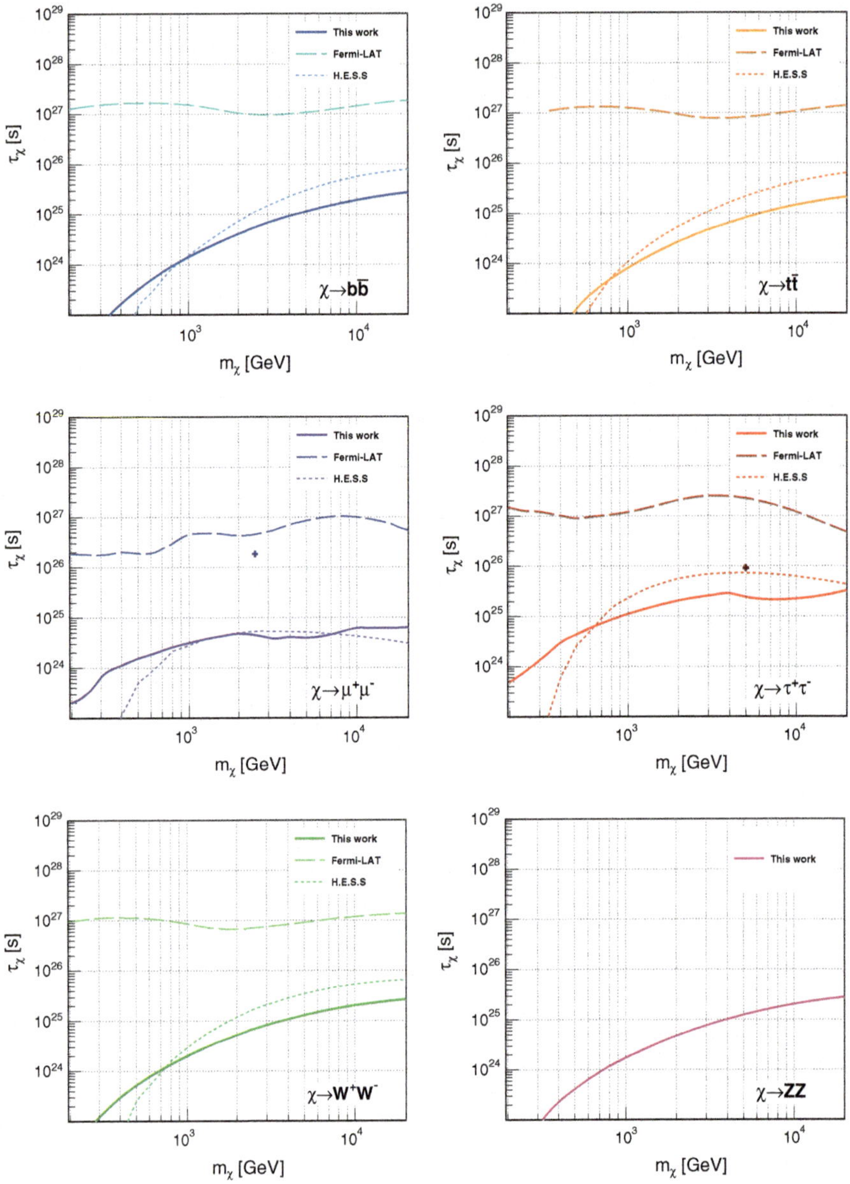

Fig. 5.39 Lower limits on τ_χ as a function of m_χ, for dark matter decay into quark-antiquark (*top row*), lepton-antilepton (*middle row*) and gauge boson pairs (*bottom row*) from the Segue 1 observations with MAGIC (*full lines*), compared to the exclusion *curves* from Fermi-LAT and H.E.S.S. data (*dashed* and *dotted lines*, respectively, [51]). For leptonic channels, $\mu^+\mu^-$ and $\tau^+\tau^-$, also shown are the best fit values to the PAMELA and Fermi cosmic-ray excess measurements (cross markers, taken from [47])

particle, $m_\chi = 20\,\text{TeV}$, and is of order of 2.6×10^{25} s. This is a factor $\sim$2 less constraining than the H.E.S.S. limit for same m_χ, and almost two orders of magnitude than the Fermi-LAT bound. This is expected, considering the choice of targets by different experiments; nevertheless, for $m_\chi < 900$ GeV, the exclusion curve from this work is actually stronger than the one from H.E.S.S.

- $\chi \rightarrow t\bar{t}$

Similarly to $b\bar{b}$, for $t\bar{t}$ channel the limit from this work is more stringent than the limit from H.E.S.S. for $m_\chi < 800\,\text{GeV}$ (Fig. 5.39). The best constraint, $\tau_\chi \sim 2\times10^{25}$ s, corresponds to $m_\chi = 20\,\text{TeV}$ and is a factor $\sim$3 and a factor $\sim$100 weaker, for the same m_χ, from the H.E.S.S. and Fermi-LAT bounds, respectively.

- $\chi \rightarrow \mu^+\mu^-$

Dark matter decay to leptonic final states has been considered as possible explanation of the measured positron fraction excesses (Fig. 2.3). The scalar dark matter decay into $\mu^+\mu^-$ pair, with $m_\chi \simeq 2500$ GeV and $\tau_\chi \simeq 1.8 \times 10^{26}$ s, has been postulated as a good fit to the observations [47]. Figure 5.39 shows the lower limits on τ_χ from the MAGIC Segue 1 observations for this channel: although $\sim$2 orders of magnitude less constraining that the Fermi-LAT bound, this limit is a factor $\sim$40 away from the best fit to the PAMELA and Fermi data (symbolized with a cross). Furthermore, τ_χ lower limits calculated in this work are comparable, or even stronger than the exclusion curve estimated from the H.E.S.S. data. The most constraining MAGIC limit corresponds to $\tau_\chi \sim 6.2 \times 10^{24}$ s, for $m_\chi = 20\,\text{TeV}$.

- $\chi \rightarrow \tau^+\tau^-$

This channel has also been considered as a good fit to the observed positron excesses, assuming a massive dark matter particle, $m_\chi = 5\,\text{TeV}$, with lifetime of $\tau_\chi = 9 \times 10^{26}$ s (marked as cross in Fig. 5.39, [47]). MAGIC lower limit on τ_χ for $\tau^+\tau^-$ final state is only a factor 4 away from this point. However, the limit derived from H.E.S.S. (up to a factor $\sim$3 stronger than the bound from this work) is in tension with it, while the Fermi-LAT bound (from 1 to 3 orders of magnitude more constraining than the MAGIC exclusion curve, depending on m_χ) already excludes this scenario. The strongest limit from this work is of order $\tau_\chi \sim 3.1 \times 10^{25}$ s, for $m_\chi = 20\,\text{TeV}$, being at the same time the most constraining result on τ_χ of all the considered final state channels from the MAGIC Segue 1 observations.

- $\chi \rightarrow W^+W^-$

The strongest lower limit on τ_χ for the W^+W^- channel is of order $\tau_\chi \sim 2.6 \times 10^{25}$ s, for $m_\chi = 20\,\text{TeV}$ (Fig. 5.36). This is a factor $\sim$50 and a factor $\sim$2 weaker than the corresponding bounds from Fermi-LAT and H.E.S.S. data (respectively). For lower dark matter particle masses ($m_\chi < 750\,\text{GeV}$), MAGIC constraints are actually stronger than those from H.E.S.S.

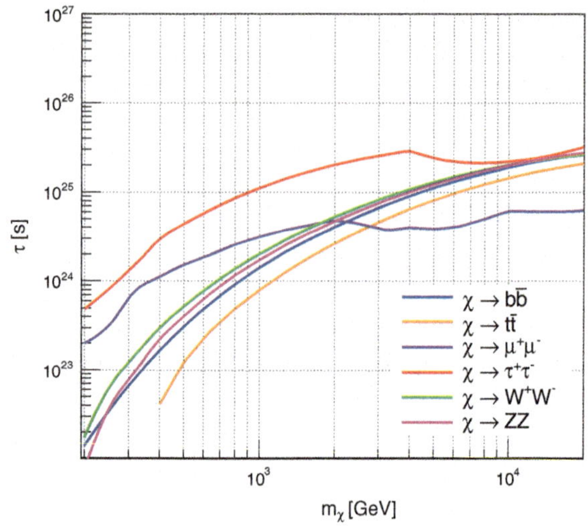

Fig. 5.40 Lower limits on τ_χ, for the considered final state channels, from Segue 1 observations with MAGIC

- $\chi \rightarrow ZZ$

Lower limits on τ_χ as a function of m_χ for the $\chi \rightarrow ZZ$ decay are similar to those for W^+W^- channel (Fig. 5.39). The strongest bound is of order $\tau_\chi \sim 2.7 \times 10^{25}$ s, for $m_\chi = 20$ TeV. No constraints from other experiments have been found for this channel.

Overall, constraints on dark matter decay into final state channels from the Segue 1 observations with MAGIC are weaker than the bounds derived from Fermi-LAT and H.E.S.S. data, which is expected given the extensions of different targets in question. Nevertheless, the limits from this work are comparable with those of H.E.S.S. from galaxy cluster, and are of order $\tau_\chi \sim (10^{25}-10^{26})$ s (Fig. 5.40).

5.4.2 Gamma-Ray Line

5.4.2.1 Annihilation

Although the direct annihilation of dark matter particles into photon(s) is highly loop suppressed ($\mathcal{O} \sim 1/\alpha^2$), the importance of the detection of the resulting gamma-ray line emission can not be overestimated: not only would a line be a firm proof of the dark matter existence, it would also reveal some information about its nature. It is why this feature has been so appealing, and many searches for a hint of it have been conducted so far, in galaxy clusters [57], Milky Way dSph satellites [56], the Galactic Center and in the Galactic Halo [53, 54, 58].

The currently strongest upper limits, presented by the Fermi-LAT, follow from 3.7 years of observations of the Galactic Center region [53] and extend from $\langle\sigma_{\text{ann}}v\rangle \leq$

$1 \times 10^{-29}\,\text{cm}^3\,\text{s}^{-1}$ at $m_\chi = 10\,\text{GeV}$ to $\langle\sigma_{\text{ann}}v\rangle \leq 8 \times 10^{-28}\,\text{cm}^3\,\text{s}^{-1}$ at $m_\chi = 200\,\text{GeV}$. At higher energies, above 500 GeV, results from 112 h of H.E.S.S. observations of the central Galactic Halo set limits of order $\langle\sigma_{\text{ann}}v\rangle \sim 3 \times 10^{-27}\,\text{cm}^3\,\text{s}^{-1}$ for $m_\chi = 1\,\text{TeV}$ [54].

In addition, it is worth mentioning the recently claimed hint of a line-like signal at ∼130 GeV in the Fermi data of the Galactic Center region [40, 55]. Although this result could not be confirmed by the Fermi-LAT Collaboration [53], potential presence of this feature has stirred the scientific community, and numerous explanations have appeared about its origin (for a review, see [59]). If the observed signal originates from direct dark matter annihilation into two photons, the WIMP particle should have a mass of $m_\chi = 129 \pm 2.4^{+7}_{-13}\,\text{GeV}$ and annihilation rate (assuming Einasto profile) of $\langle\sigma_{\text{ann}}v\rangle_{\gamma\gamma} = (1.27 \pm 0.32^{+0.18}_{-0.28}) \times 10^{-27}\,\text{cm}^3\,\text{s}^{-1}$. For the $Z\gamma$ final state, the feature comes from the dark matter particle of $m_\chi = 144.2 \pm 2.2^{+6}_{-12}\,\text{GeV}$, with $\langle\sigma_{\text{ann}}v\rangle_{Z\gamma} = (3.14 \pm 0.79^{+0.40}_{-0.60}) \times 10^{-27}\,\text{cm}^3\,\text{s}^{-1}$.

Search for line-like features in this work has been done assuming a 100 % branching ratio into $\gamma\gamma$ and $Z\gamma$ final states. No signal has been found. The outcome of the analysis is then transformed into upper limits on $\langle\sigma_{\text{ann}}v\rangle$. First, those limits are compared with the expectations for the no-signal scenario, as well as for the 1σ and 2σ signal significances, and then with the existing bounds from the other gamma-ray observatories.

- $\chi\chi \to \gamma\gamma$

Figure 5.41 shows the upper limits on $\langle\sigma_{\text{ann}}v\rangle$for direct annihilation of dark matter particles into two photons. For the considered m_χ range, the constraints are set between $\langle\sigma_{\text{ann}}v\rangle \sim (4 \times 10^{-26} – 10^{-24}\,\text{cm}^3\,\text{s}^{-1})$. Dashed line and shaded gray and

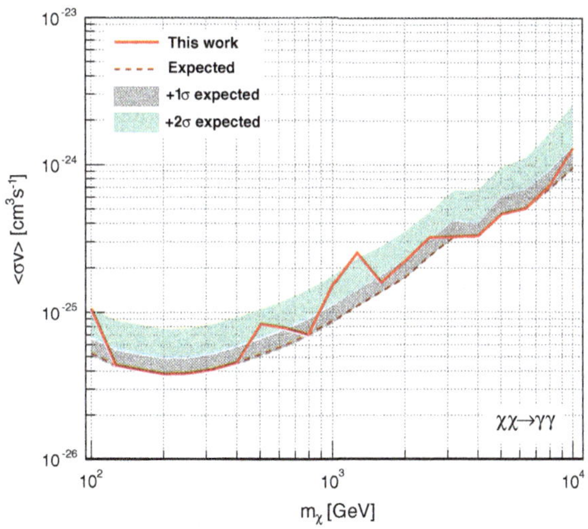

Fig. 5.41 Upper limits on $\langle\sigma_{\text{ann}}v\rangle$ for annihilation into two photons as a function of m_χ, from the Segue 1 observations with MAGIC (*full line*) and as expected for the case of no signal (*dashed line*), or for the signal with 1σ or 2σ significance (*gray* and *light blue shaded areas*, respectively)

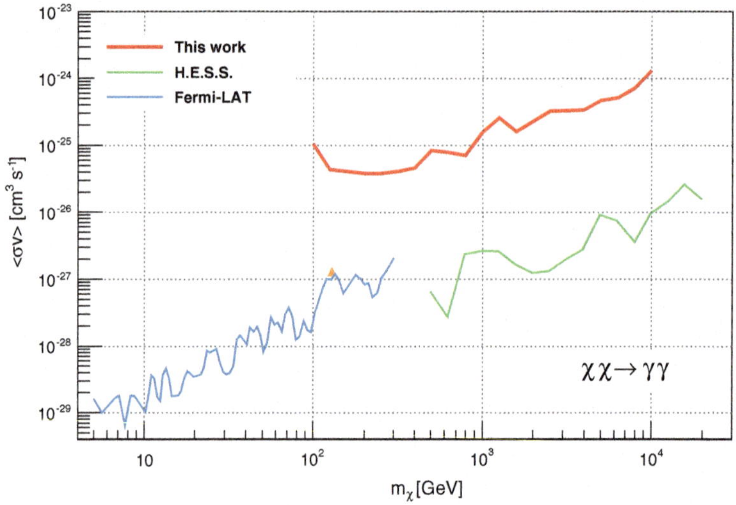

Fig. 5.42 Upper limits on $\langle\sigma_{\mathrm{ann}} v\rangle$ for dark matter annihilation into two photons, from this work (*red line*), compared with the exclusion *curves* from the Galactic Center region observations from Fermi-LAT [53] (3.7 years, *blue line*) and H.E.S.S [54] (112 h, *green line*). Also shown is the $\langle\sigma_{\mathrm{ann}} v\rangle$ value corresponding to the 130 GeV gamma-ray line [55] (*orange triangle*)

blue areas show the expectations for no signal, 1σ and 2σ signal significance scenarios.

With respect to the results from the Galactic Center region, bounds from the dSphs observations (in general) are not very competitive, as the high astrophysical factor J of the former ones always provides an advantage. Still, as already mentioned, there are significant uncertainties in the J estimation, and besides, it is very interesting to see how limits from different targets compare. Figure 5.42 shows the limit from this work versus the strongest constraints from the Fermi-LAT [53] and H.E.S.S. line searches [54]. For $m_\chi = 200\,\mathrm{GeV}$, MAGIC result is about an order of magnitude away from that of Fermi-LAT, while the difference with the H.E.S.S. limit is almost two orders of magnitude. Again, considering the targets in question for these observations, these relations are as expected. Also shown is the $\langle\sigma_{\mathrm{ann}} v\rangle$ value corresponding to the $\chi\chi \to \gamma\gamma$ explanation for the 130 GeV feature (triangle marker in Fig. 5.42). MAGIC bound is a factor $\sim$30 away from testing this claim.

Additionally, constraints from this work are compared to the results of the Fermi-LAT joint analysis of 10 dSphs [22], that have been translated into exclusion curve for the case of direct annihilation into two photons [56]. As shown in Fig. 5.43, Fermi-LAT and MAGIC bounds complement each other, with MAGIC limit becoming more stringent for $m_\chi > 130$ GeV. Thus, when the dark matter target sources are of the same kind, these two observatories are complementary.

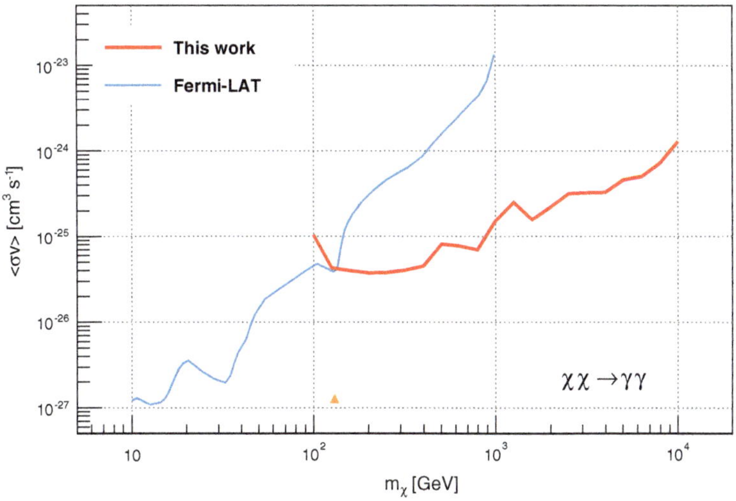

Fig. 5.43 Upper limits on $\langle\sigma_{ann}v\rangle$ for dark matter annihilation into two photons, from this work (*red line*), compared with the exclusion *curve* from the 2 years of Fermi-LAT dSphs observations [56] (*blue line*). Also shown is the $\langle\sigma_{ann}v\rangle$ value corresponding to the 130 GeV gamma-ray line [55] (*orange triangle*)

Fig. 5.44 Upper limits on $\langle\sigma_{ann}v\rangle$ for dark matter annihilation into a Z boson and a photon, from this work (*red line*), compared with the exclusion curve from 2 years of the Galactic Center region observations with Fermi-LAT [58] (*blue line*)

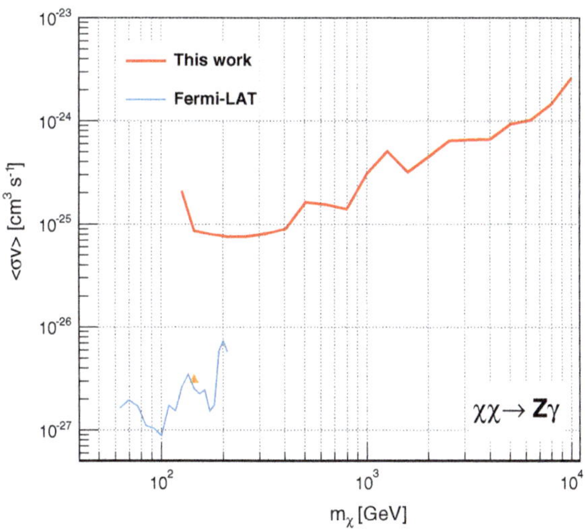

- $\chi\chi \to Z\gamma$

Line from direct dark matter annihilation into a photon and a Z boson is somewhat smeared due to the finite width of the Z boson, and it is centered at energy of $E_\gamma = m_\chi\left(1 - m_Z^2/4m_\chi^2\right)$. Figure 5.44 shows the upper limits on $\langle\sigma_{ann}v\rangle$ for $\chi\chi \to Z\gamma$ annihilation, calculated from the Segue 1 observations with MAGIC, in comparison with the exclusion curve obtained from the 2 years of Fermi-LAT measurements of the Galactic Center region [58]. The most constraining bound from this work corresponds to $\langle\sigma_{\text{ann}}v\rangle \sim 7\times10^{-26}$ cm^3 s^{-1}, for $m_\chi \sim 250$ GeV. Also shown is the $\langle\sigma_{\text{ann}}v\rangle$ estimated for the $Z\gamma$ explanation of the line-like feature at 130 GeV; MAGIC upper limit is a factor ~30 away from this value.

5.4.2.2 Decay

If the dark matter particle is a gravitino in R-parity breaking vacua, and its life time is of order $\tau_\chi \sim 10^{27}$ s or larger [49], it can decay to a photon and a neutrino, producing one monochromatic gamma-ray line at energy of $E_\gamma \simeq m_\chi/2$. Search for gamma-ray line of such origin has been conducted by the Fermi-LAT, in 2-year of observations of the Galactic Center region [58]. The derived bounds, for dark matter particle mass in the 14 to 400 GeV range, are of order $\tau_\chi \sim (4-30)\times10^{29}$ s.

Figure 5.45 shows the results from this work compared to those Fermi-LAT lower limits on τ_χ. Although the considered m_χ range extends well beyond the energies required for decay into W or Z bosons (that would consequently fragment into photons with continuous spectrum), here is considered only the monochromatic emission. MAGIC limits are almost three orders of magnitude less constraining than those of Fermi-LAT for complementary energy ranges; the strongest bound from this work is of order $\tau_\chi \sim 2\times10^{26}$ s for dark matter particle of mass 1.5 TeV.

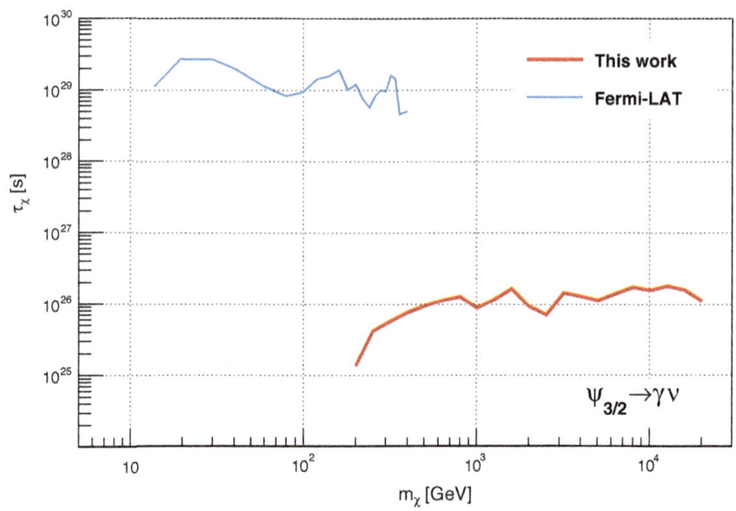

Fig. 5.45 Lower limits on τ_χ for dark matter decay into neutrino and photon, from this work (*red line*), compared with the exclusion *curve* from 2 years of the Galactic Center region observations with Fermi-LAT [58] (*blue line*)

The case of the decay of scalar dark matter into two photons is not considered, but it is trivial to derive the τ_χ lower limits for that scenario: the gamma-ray signal would be the same as for the $\gamma\nu$ channel, only twice as strong.

5.4.3 Virtual Internal Bremsstrahlung

The contribution of VIB photons to the gamma-ray spectrum has a positive impact on the dark matter detection prospects, as not only it represents a feature that can be rather pronounced (and even dominate over the monochromatic gamma-ray lines, given the energy resolution of current detectors), but it can also significantly enhance the expected flux at the highest energies (Sects. 2.3.2.3 and 5.3.5).

Assuming a fermionic dark matter particle, that couples to muons or tau leptons via Yukawa interactions with the scalar η, helicity suppression in the s-wave contribution to the $\langle\sigma_{\rm ann}v\rangle$ can be lifted by emission of a VIB photon. Depending on the mass-splitting parameter μ, that is defined as the squared ratio of the masses of the scalar η and dark matter particle ($\mu = (m_\eta/m_\chi)^2$), the VIB contribution to the spectrum is more or less pronounced (Fig. 5.24). Here is considered the 3-body annihilation into $\mu^+\mu^-(\gamma)$ and $\tau^+\tau^-(\gamma)$ channels, with FSR included in the continuous part of the spectrum.

Figure 5.46 shows the $\langle\sigma_{\rm ann}v\rangle$ upper limits for the $\mu^+\mu^-(\gamma)$ channel, estimated from the Segue 1 observations with MAGIC, for different values of the mass splitting parameter, $\mu = 1.05$, 1.50 and 2.00, chosen so that the VIB contribution is significant with respect to the continuous one. The exclusion curves are rather similar, which can be understood by comparing their corresponding spectral shapes after the convolution with the MAGIC response function (Fig. 5.24). Still, it can be noticed that the most degenerate case, $\mu = 1.05$, provides the strongest limit, of $\langle\sigma_{\rm ann}v\rangle \sim 9\times10^{-26}$ cm^3 s^{-1} for $m_\chi \sim 250$ GeV. Also shown are the predictions for the null-hypothesis, as well as the expectations for 1σ and 2σ significant signal.

The relevance of the VIB contribution becomes more apparent from Fig. 5.47, where the detection prospects are compared for annihilation into the muon pair with and without the emission of the VIB photons (Sect. 5.4.1.1, Fig. 5.31). In this particular case, the presence of VIB contribution means almost two orders of magnitude more stringent limits. For the $\tau^+\tau^-(\gamma)$ channel (Fig. 5.48), the gain from VIB presence is especially significant for lower masses m_χ—more than an order of magnitude stronger $\langle\sigma_{\rm ann}v\rangle$ bounds, while for $m_\chi > 1$ TeV, depending on μ value, the improvement is a factor 5 or more.

Limits for $\mu^+\mu^-(\gamma)$ and $\tau^+\tau^-(\gamma)$ channels are very similar (Figs. 5.47 and 5.48). The reason is that, for the considered range of values of μ, the gamma-ray spectrum is dominated by the VIB contribution in both cases, as seen in Figs. 2.8 and 5.24. In addition, after the convolution with the response function of the telescopes, these spectra become very similar (Fig. 5.24); thus the comparable limits (although, slightly more constraining for $\tau^+\tau^-(\gamma)$ channel). The strongest $\langle\sigma_{\rm ann}v\rangle$ limit for $\tau^+\tau^-(\gamma)$ channels is of order of 8.7×10^{-26} cm^3 s^{-1}, for $\mu = 1.05$ and $m_\chi = 250$ GeV.

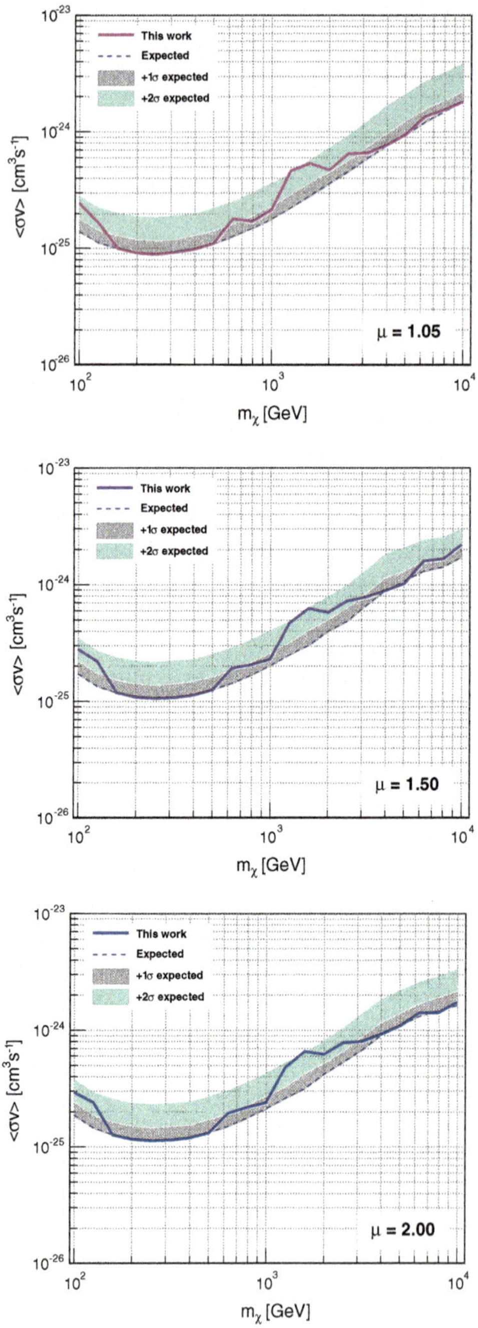

Fig. 5.46 Upper limits on $\langle\sigma_{\rm ann} v\rangle$ for $\mu^+\mu^-(\gamma)$ channel as a function of m_χ, from the Segue 1 observations with MAGIC (*full line*) and as expected for the case of no signal (*dashed line*), or for the signal with 1σ or 2σ significance (*gray* and *light blue shaded areas*, respectively). The value of the mass splitting parameter μ is 1.05 (*top*), 1.50 (*middle*) and 2.00 (*bottom*)

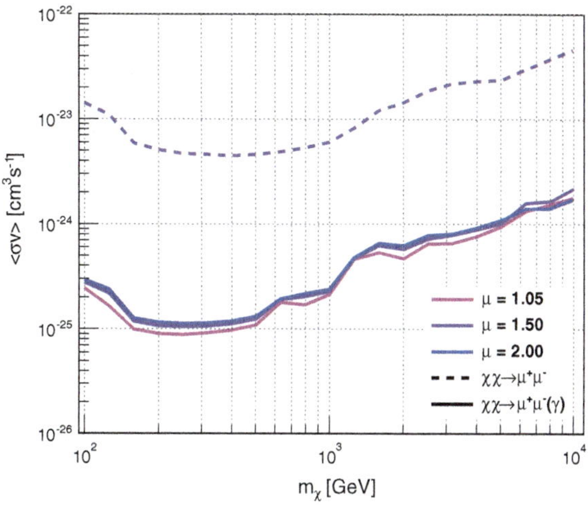

Fig. 5.47 Upper limits on $\langle\sigma_{\text{ann}}v\rangle$ for $\mu^+\mu^-(\gamma)$ channel, as a function of m_χ, for different values of the mass splitting parameter μ: 1.05, 1.50 and 2.00 (*full pink, violet* and *blue lines*, respectively). Also shown is the exclusion *curve* for the annihilation without the VIB contribution (*dashed line*)

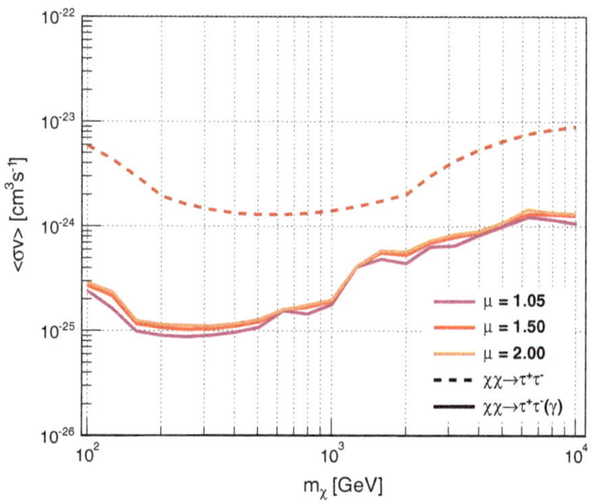

Fig. 5.48 Upper limits on $\langle\sigma_{\text{ann}}v\rangle$ for $\tau^+\tau^-(\gamma)$ channel, as a function of m_χ, for different values of the mass splitting parameter μ: 1.05, 1.50 and 2.00 (*full pink, red* and *orange lines*, respectively). Also shown is the exclusion *curve* for the annihilation without the VIB contribution (*dashed line*)

5.4.4 Gamma-Ray Boxes

Gamma-ray boxes—the sharp spectral features that are expected to originate from cascade annihilation or decay of dark matter particles (into intermediated scalars ϕ that in turn decay into photons)—would be, if observed, another conclusive proof of dark matter existence.

This work considers the case of dark matter annihilation resulting in four photons[4] ($\chi\chi \to \phi\phi \to \gamma\gamma\gamma\gamma$). Figure 5.49 shows the $\langle\sigma_{\text{ann}}v\rangle$ exclusion curves as a function of m_χ, calculated from the Segue 1 observations with MAGIC, assuming extreme degeneracies, when $m_\phi/m_\chi = 0.1$ and $m_\phi/m_\chi = 0.99$. In both cases, the strongest

[4] In principle, the intermediate scalar can also decay into photon and gauge boson, or two bosons.

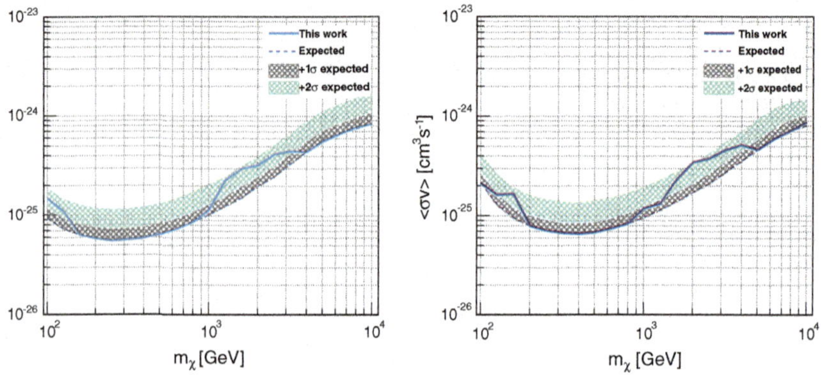

Fig. 5.49 Upper limits on $\langle\sigma_{\rm ann}v\rangle$ for wide- and narrow-box scenarios ($m_\phi/m_\chi = 0.1$ and 0.99, respectively), as a function of m_χ, from the Segue 1 observations with MAGIC (*full lines*), and as expected for the case of no signal (*dashed lines*), or for the signal with 1σ or 2σ significance (*gray* and *light blue shaded areas*, respectively)

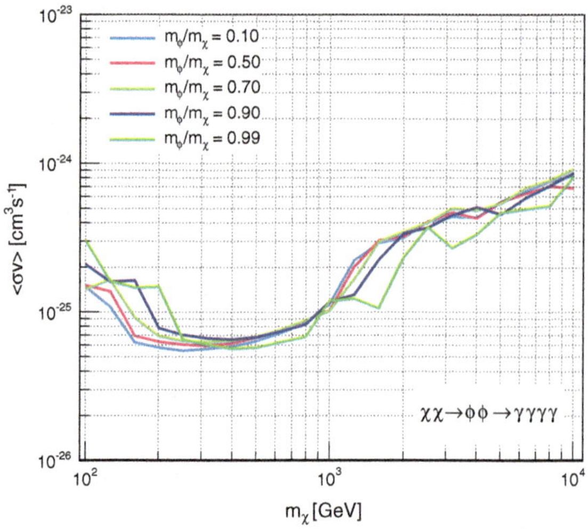

Fig. 5.50 Upper limits on $\langle\sigma_{\rm ann}v\rangle$, as a function of m_χ, from Segue 1 observations with MAGIC and for different ratios of scalar and dark matter particle masses: $m_\phi/m_\chi = 0.1, 0.5, 0.7, 0.9$ and 0.99 (*blue*, *pink*, *light green*, *violet* and *green lines*, respectively)

constraints are similar, of order $\langle\sigma_{\rm ann}v\rangle \sim (5-6)\times 10^{-25}\,{\rm cm}^3\,{\rm s}^{-1}$. These exclusion curves are compared with the expectations from toy MC, assuming no signal in the data, or signal with 1σ or 2σ significance.

Figure 5.50 represents upper limits on $\langle\sigma_{\rm ann}v\rangle$ for various values of m_ϕ/m_χ. As it can be seen, with exception of the most narrow box scenario, all of the constraints are essentially the same, and only a factor few weaker that the most degenerate configuration. This is understood given that the wide boxes compensate the dimmer amplitudes (with respect to the $m_\phi \approx m_\chi$ cases) by extending to higher energies, where sensitivity of the telescopes is better.

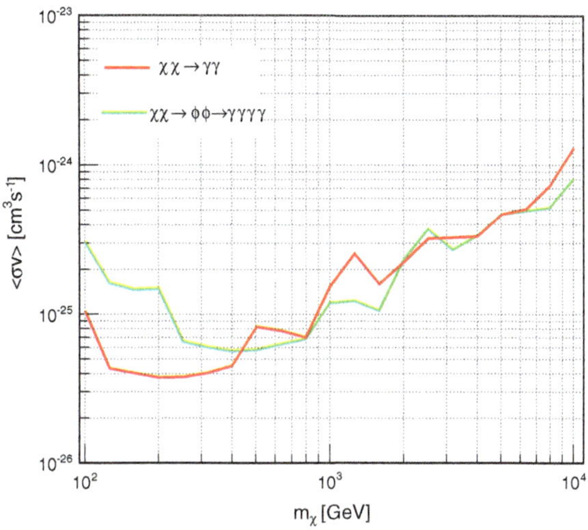

Fig. 5.51 Comparison of the upper limits on $\langle\sigma_{\rm ann}v\rangle$, as a function of m_χ, from Segue 1 observations with MAGIC, for a narrow-box scenario ($m_\phi/m_\chi = 0.99$, *green line*) and for a monochromatic gamma-ray line (*red line*)

For a more general view on the importance of box-shaped features, Fig. 5.51 shows the upper limits on $\langle\sigma_{\rm ann}v\rangle$ from the most degenerate box model ($m_\phi/m_\chi = 0.99$) and from the line searches conducted in this work (Fig. 5.41, Sect. 5.4.2.1). It can be seen that the obtained bounds for both models are of same order of magnitude, although the direct comparison between the two exclusion curves is not immediate: line is centered at $E_\gamma = m_\chi$ and is normalized for 2 photons, while the box-shaped feature is centered at $E_\gamma = m_\chi/2$ and it is normalized for 4 photons. This is reflected as a shift in x and y coordinates of the exclusion plots.

The limits obtained here are compared with the recent estimations made for the Fermi-LAT and H.E.S.S. observatories [60], for scenarios with very narrow and very wide boxes ($m_\phi/m_\chi = 0.999$ and $m_\phi/m_\chi = 0.1$, respectively), assuming that intermediate particle is an axion that decays into two photons, and that the measured data (used to derive the bounds) are entirely from the background emission (Fig. 5.52). Like in the case of lines, observations of the Galactic Center regions are more constraining than those of the dSph galaxies. For both of the considered box widths, the $\langle\sigma_{\rm ann}v\rangle$ upper limits from MAGIC measurements are about an order of magnitude away from the H.E.S.S. constraints, for the greater part of the considered m_χ range. As for the Fermi-LAT limits, in the overlapping mass domain, the results from this work are a factor few weaker.

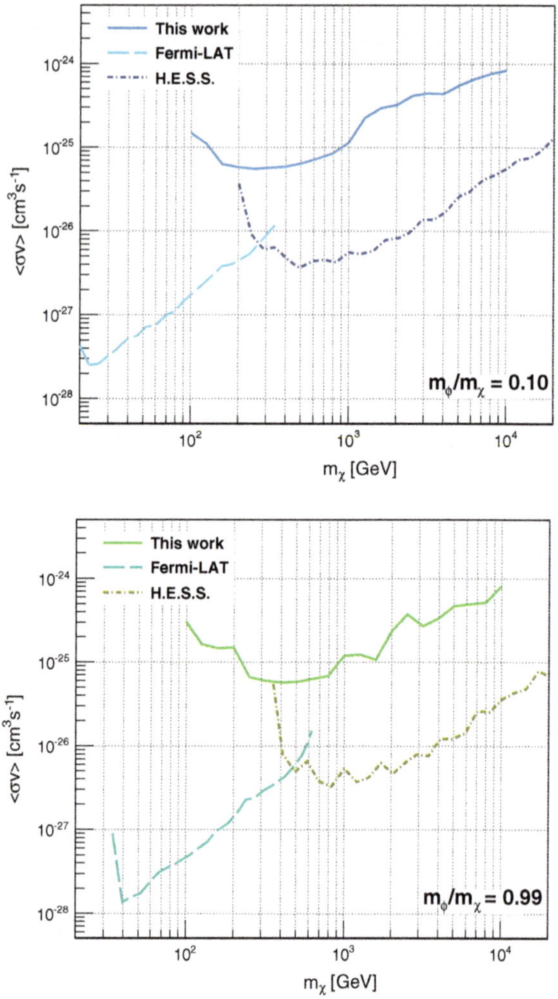

Fig. 5.52 Upper limits on $\langle\sigma_{\rm ann} v\rangle$ for the wide ($m_\phi/m_\chi = 0.1$, *top*) and narrow box scenario ($m_\phi/m_\chi = 0.99$, *bottom*), from this work (*full lines*) and from estimations made for the Fermi-LAT and H.E.S.S. observatories (*dashed* and *dash-dotted lines*, respectively). The latter are adapted from [60], for Majorana dark matter particle

References

1. L.E. Strigari, Galactic searches for dark matter (2012), arXiv:1211.7090
2. V. Springel et al., The Aquarius project: the subhaloes of galactic haloes. Mon. Not. Roy. Astron. Soc. **391**, 1685–1711 (2008). arXiv:0809.0898
3. B. Moore et al., Cold collapse and the core catastrophe. Mon. Not. Roy. Astron. Soc. **310**, 1147–1151 (1999). arXiv:astro-ph/9903164
4. G. Gilmore et al., The observed properties of dark matter on small spatial scales. Astrophys. J. **663**, 948–959 (2007). arXiv:astro-ph/0703308
5. H. Shapley, A stellar system of a newtype. Harvard Coll. Obs. Bull. **908**, 1–11 (1938)
6. J. Diemand, M. Kuhlen, P. Madau, Dark matter substructure and gamma-ray annihilation in the Milky Way Halo. Astrophys. J. **657**, 262–270 (2007). arXiv:astro-ph/0611370

7. The Sloan Digital Sky Survey Collaboration, Sloan Digital Sky Survey home page (2013), http://www.sdss.org
8. M.G. Walker, M. Mateo, E.W. Olszewski, Stellar velocities in the Carina, Fornax, Sculptor, and Sextans dSph galaxies: data from the Magellan/MMFS Survey. Astron. J. **137**, 3100–3108 (2009). arXiv:0811.0118
9. L. Strigari et al., Redefining the missing satellites problem. Astrophys. J. **669**, 676–683 (2007). arXiv:0704.1817
10. L. Strigari et al., A common mass scale for satellite galaxies of the Milky Way. Nature **454**, 1096–1097 (2008). arXiv:0808.3772
11. J. Klimentowski et al., Mass modelling of dwarf spheroidal galaxies: the effect of unbound stars from tidal tails and the Milky Way. Mon. Not. Roy. Astron. Soc. **378**, 353–368 (2007). arXiv:astro-ph/0611296
12. J.D. Simon, M. Geha, The kinematics of the ultra-faint Milky Way satellites: solving the missing satellite problem. Astrophys. J. **670**, 313–331 (2007). arXiv:0706.0516
13. B.J. Pritzl, K.A. Venn, M. Irwin, A comparison of elemental abundance ratios in globular clusters, field stars, and dwarf spheroidal galaxies. Astrophys. J. **130**, 2140–2165 (2005). arXiv:astro-ph/0506238
14. D. Harbeck et al., Population gradients in Local Group dwarf spheroidals. Astron. J. **122**, 3092–3105 (2001). arXiv:astro-ph/0109121
15. M. Mateo et al., The Carina dwarf spheroidal galaxy—how dark is it? Astron. J. **105**, 510–526 (1993)
16. J. Simon et al., A complete spectroscopic survey of the Milky Way satellite Segue 1: the darkest galaxy. Astrophys. J. **733**, 46–66 (2011). arXiv:1007.4198
17. C.J. Hogan, J.J. Dalcanton, New dark matter physics: clues from halo structure. Phys. Rev. D **63**, 063511 (2000). arXiv:astro-ph/0002330
18. M. Mateo, Dwarf galaxies of the Local Group. Ann. Rev. A&A **36**, 435–506 (1998). arXiv:astro-ph/9810070
19. N. Evans, F. Ferrer, S. Sarkar, A Baedecker for the dark matter annihilation signal. Phys. Rev. D **69**, 123501 (2004). arXiv:astro-ph/0311145
20. J. Jardel, K. Gebhardt, The dark matter density profile of the Fornax dwarf. Astrophys. J. **746**, 89–97 (2012). arXiv:1112.0319
21. L.E. Strigari et al., Precise constraints on the dark matter content of Milky Way dwarf galaxies for gamma-ray experiments. Phys. Rev. D **75**, 083526 (2007). arXiv:astro-ph/0611925
22. M. Ackermann et al., Constraining dark matter models from a combined analysis of Milky Way satellites with the Fermi Large Area Telescope. Phys. Rev. Lett. **107**, 241302 (2011). arXiv:1108.3546
23. J. Albert et al., Upper limit for gamma-ray emission above 140 GeV from the dwarf spheroidal galaxy Draco. Astrophys. J. **679**, 428–431 (2008). arXiv:0711.2574
24. E. Aliu et al., MAGIC upper limits on the VHE gamma-ray emission from the satellite galaxy Willman 1. Astrophys. J. **697**, 1299–1304 (2009). arXiv:0810.3561
25. J. Aleksić et al., Searches for dark matter annihilation signatures in the Segue 1 satellite galaxy with the MAGIC-I telescope. JCAP **06**, 035 (2011), arXiv:1103.0477
26. V.A. Acciari et al., VERITAS search for VHE gamma-ray emission from dwarf spheroidal galaxies. Astrophys. J. **720**, 1174–1180 (2010). arXiv:1006.5955
27. E. Aliu et al., VERITAS deep observations of the dwarf spheroidal galaxy Segue 1. Phys. Rev. D **85**, 062001 (2012). arXiv:1202.2144
28. F. Aharonian et al., A search for a dark matter annihilation signal towards the Canis Major overdensity with H.E.S.S., Astrophys. J. **691**, 175–181 (2009), arXiv:0809.3894
29. F. Aharonian, Observations of the Sagittarius dwarf galaxy by the H.E.S.S. experiment and search for a dark matter signal, Astropart. Phys. **29**, 55–62 (2008), Erratum-ibid.33: 274–275 (2010), arXiv:0711.2369
30. A. Abramowski, H.E.S.S. constraints on dark matter annihilations towards the Sculptor and Carina dwarf galaxies. Astropart. Phys. **34**, 608–616 (2011). arXiv:1012.5602

31. The Sloan Digital Sky Survey Collaboration, SEGUE: Sloan Extension for Galactic Understanding and Exploration, http://www.sdss.org/segue/aboutsegue (2013)
32. V. Belokurov et al., Cats and dogs, hair and a hero: a quintet of new Milky Way companions. Astrophys. J. **654**, 897–906 (2007). arXiv:astro-ph/0608448
33. Keck Observatory, DEIMOS home page, (2013), http://www2.keck.hawaii.edu/inst/deimos
34. M. Geha et al., The least luminous galaxy: spectroscopy of the Milky Way satellite Segue 1. Astrophys. J. **692**, 1464–1475 (2009). arXiv:0809.2781
35. M. Niederste-Ostholt et al., The origin of Segue 1. Mon. Not. Roy. Astron. Soc. **398**, 1771–1781 (2009). arXiv:0906.3669
36. W. A. Rolke, A. M. López, J. Conrad, Limits and confidence intervals in the presence of nuisance parameters. Nucl. Instrum. Meth. A **551**, 493–503 (2005), physics/0403059
37. J. Aleksić et al., Performance of the MAGIC stereo system obtained with the Crab Nebula data. Astropart. Phys. **35**, 435–448 (2012), arXiv:1108.1477
38. F. James, MINUIT. Function minimization and error analysis, reference manual version 94.1, CERN Program Library Long Writeup D 506, CERN, Geneva, Switzerland (1994)
39. J. Cembranos et al., Photon spectra from WIMP annihilation. Phys. Rev. D **83**, 083507 (2011). arXiv:1009.4936
40. T. Bringmann et al., Fermi LAT search for internal bremsstrahlung signatures from dark matter annihilation. JCAP **07**, 054 (2012). arXiv:1203.1312
41. J.F. Navarro et al., The diversity and similarity of simulated cold dark matter haloes. Mon. Not. Roy. Astron. Soc. **402**, 21–34 (2010). arXiv:0810.1522
42. R. Essig et al., Indirect dark matter detection limits from the ultra-faint Milky Way satellite Segue 1. Phys. Rev. D **82**, 123503 (2010). arXiv:1007.4199
43. The ROOT Team, T Minuit, (2013), http://root.cern.ch/root/html/TMinuit.html
44. A. Abramowski et al., Search for a dark matter annihilation signal from the Galactic Center Halo with H.E.S.S. Phys. Rev. Lett. **106**, 161301 (2011). arXiv:1103.3266
45. J. Aleksić, J. Rico, M.Martinez, Optimized analysis method for indirect dark matter searches with imaging air Cherenkov telescopes, JCAP **10**, 032 (2012), arXiv:1209.5589
46. The VERITAS Collaboration, VERITAS Specifications, (2013), http://veritas.sao.arizona.edu/about-veritas-mainmenu-81/veritas-speci_cations-mainmenu-111
47. A. Ibarra, D. Tran, C. Weniger, Decaying dark matter in light of the PAMELA and Fermi LAT data. JCAP **01**, 009 (2010). arXiv:0906.1571
48. A. Ibarra, D. Tran, Decaying dark matter and the PAMELA anomaly. JCAP **02**, 021 (2009). arXiv:0811.1555
49. G. Bertone et al., Gamma-rays from decaying dark matter. JCAP **11**, 003 (2012). arXiv:0709.2299
50. M. Ackermann, Fermi-LAT Results on the intensity and the origin of the extragalactic gamma-ray background. *Talk at the TeVPA Conference*, Stockholm, Sweden (2011)
51. M. Cirelli et al., Gamma ray constraints on decaying dark matter. Phys. Rev. D **86**, 083506 (2012). arXiv:1205.5283
52. A. Abramowski et al., Search for dark matter annihilation signals from the Fornax galaxy cluster with H.E.S.S. Astrophys. J. **750**, 123 (2012). arXiv:1202.5494
53. M. Ackermann et al., Search for gamma-ray spectral lines with the Fermi Large Area Telescope and dark matter implications (2013), arXiv:1305.5597
54. A. Abramowski et al., Search for photon line-like signatures from dark matter annihilations with H.E.S.S. Phys. Rev. Lett. **110**, 041301 (2013). arXiv:1301.1173
55. C. Weniger, A tentative gamma-ray line from dark matter annihilation at the Fermi Large Area Telescope. JCAP **08**, 007 (2012). arXiv:1204.2797
56. A. Geringer-Sameth, S.M. Koushiappas, Dark matter line search using a joint analysis of dwarf galaxies with the Fermi Gamma-ray Space Telescope. Phys. Rev. D **86**, 021302 (2012). arXiv:1206.0796
57. X. Huanga et al., Constraints on the dark matter annihilation scenario of Fermi 130 GeV gamma-ray line emission by continuous gamma rays, Milky Way halo, galaxy clusters and dwarf galaxies observations. JCAP **11**, 048 (2012). arXiv:1208.0267

58. M. Ackermann et al., Fermi LAT search for dark matter in gamma-ray lines and the inclusive photon spectrum. Phys. Rev. D **86**, 022002 (2012). arXiv:1205.2739
59. T. Bringmann, C. Weniger, Gamma ray signals from dark matter: concepts, status and prospects. Dark Universe **1**, 194–217 (2012). arXiv:1208.5481
60. A. Ibarra et al., Gamma-ray boxes from axion-mediated dark matter. JCAP **05**, 016 (2013). arXiv:1303.6632

Chapter 6
Future Prospects

Despite the earnest efforts and competitive results, like those presented in the previous chapter, the current generation of IACTs can only go so far in search for dark matter. At one point, accumulation of additional hours no longer increases the sensitivity, and the energy and angular resolution of the telescopes cannot be improved any further. Therefore, a question naturally arises about the next steps in the indirect searches at the VHE domain.

Fortunately, progress towards the future of Cherenkov astronomy is already well under way: the Cherenkov Telescope Array (CTA, [1]) will mark the new generation of IACTs, covering the energy range from few tens of GeV to more than 100 TeV, and with an order of magnitude better sensitivity than the current instruments. As such, it will not only be able to detect more than thousand sources, but also have a chance for deeper insight into the extreme phenomena of our Universe.

This chapter briefly introduces the main characteristic of the CTA and its prospects for dark matter detection.

6.1 Cherenkov Telescope Array

Following the great success of the current Cherenkov telescopes and their major contributions to the gamma-ray astronomy, advances have been made towards the construction of the CTA—the IACT of next generation. Thanks to the significant maturity the technique has acquired over the last decades, CTA will be able to count on superior performance, and thus address some issues that are not well within the reach of the present instruments: the origin of cosmic rays, the variety and properties of particle accelerators, the ultimate nature of matter and physics beyond the SM are just some of the questions that will be investigated and that might finally be understood [2].

J. Aleksić, *Optimized Dark Matter Searches in Deep Observations of Segue 1 with MAGIC*, Springer Theses, DOI 10.1007/978-3-319-23123-5_6

Fig. 6.1 Artist's impression of the CTA, with telescopes of different sizes visible. *Credit* G. Pérez/IAC/SMM

The CTA Consortium has developed to a global initiative, supported by more than thousand scientists from 27 countries. The aim is to build a detector that provides a full-sky coverage at VHE with unprecedented sensitivity, angular and timing resolution. Furthermore, CTA will operate as a proposal-driven open observatory, with transparent access to data, analysis tools and users training. Currently in the preparatory phase, CTA should become fully operational in 2019/2020.

CTA is planed to consist of two sites, one in the Southern and one in the Northern hemisphere. The Southern array will, given its privileged position, focus on the wealth of sources in the central Galactic region. Its 'baseline' configuration will consist of three types of telescopes of different mirror sizes (Fig. 6.1): the large, 24 m class telescopes, with moderate FoV (4°–5°) and very low energy threshold of ~20 GeV; the medium size telescopes, with 10–12 m in diameter and 6°–8° FoV, covering energy range between 100 GeV and 1 TeV; and the numerous small size telescopes, with mirrors of only 4–6 m in diameter, FoV around 10°, and covering the higher

Table 6.1 Compilation of performance goals for the CTA observatories [2], in the Southern (S) and Northern (N) hemispheres

	CTA	MAGIC
Coll. Area [m^2]	>10^4 (at 1 TeV) >10^6 (S, at 10 TeV) >5 × 10^5 (N, at 10 TeV)	~7 × 10^4 (at 1 TeV)
Ang. Res. (68 %) [°]	0.1 (at 100 GeV) 0.05 (>1 TeV)	0.17 (at 100 GeV) 0.08 (at 1 TeV)
En. Res. [%]	≤25 (at 50 GeV) ≤10 (>1 TeV)	19 (at 100 GeV) 15 (>1 TeV)
FoV diameter [°]	5 (at 100 GeV) 8 (at 1 TeV) 10 (>10 TeV)	3.5

Also shown are the corresponding values for the MAGIC Telescopes [3]

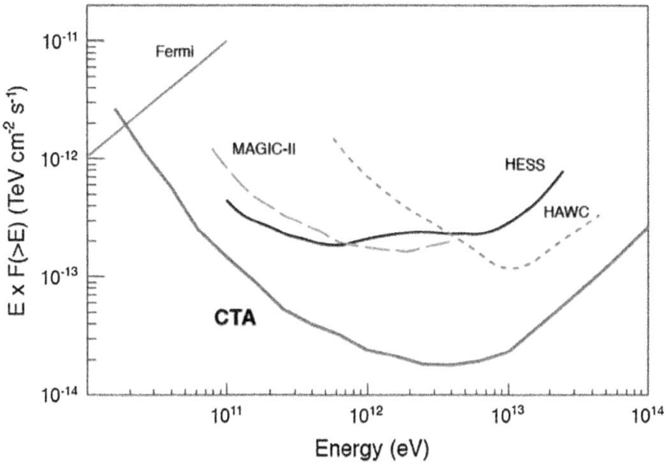

Fig. 6.2 The integral sensitivity for CTA from MC simulations, together with the sensitivity in comparable conditions (50 h for IACTs, 1 year for Fermi-LAT and HAWC) for other gamma-ray observatories. *Credit* the CTA Consortium

energy range (above 10 TeV). The Northern site will focus on lower energies, and it will only employ large and medium telescopes. Locations and layouts for both the Southern and Northern arrays are still being decided.

Table 6.1 summarizes some of the performance goals of CTA. For comparison, the corresponding MAGIC properties are listed as well. Figure 6.2 illustrates the predictions for the CTA integral sensitivity, and the improvement it represents with respect to current gamma-ray observatories.

6.2 Dark Matter Searches with CTA

Discovering physics beyond the SM and revealing the nature of dark matter are some of the key scientific objectives of CTA. Towards this goal, CTA will have significant advantages with respect to the current IACTs: the extended energy range will allow searches for lighter WIMPs; the overall improved sensitivity will increase the chances of detection, or even identification of dark matter; the increased FoV, together with the improved angular resolution, will allow for much more efficient searches for extended sources and spatial anisotropies; lastly, better energy resolution could lead to recognition of spectral features characteristic for gamma-ray emission produced in dark matter annihilation or decay. For a detailed review on CTA prospects for dark matter searches, refer to [7].

In addition, it is noteworthy, in the context of future indirect dark matter searches, to mention an initiative, led by the CTA US group, to increase the CTA sensitivity for dark matter signatures by adding an extension to the Southern site baseline

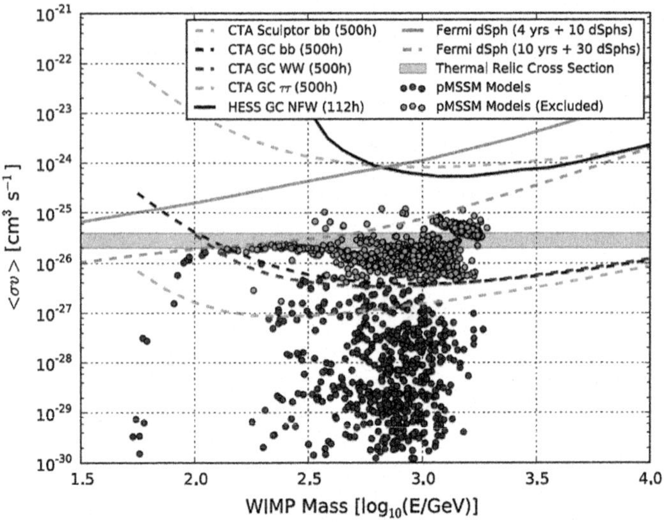

Fig. 6.3 Comparison of current (*solid lines*) and projected (*dashed lines*) limits on $\langle\sigma_{\text{ann}} v\rangle$ from different gamma-ray searches as a function of m_χ. Projected limits for CTA are shown for annihilation to several final state channels and 500 h of observations of dSph Sculptor and a 500 h observation of the Galactic Center. *Filled circles* represent pMSSM models. Models indicated in *red* would be excluded by the CTA 95 % c.l. upper limit from a 500 h observation of the Galactic Center. Taken from [4]. (Color in online)

configuration [4]. With 36 additional medium size telescopes, of very fine pixelization, the angular resolution is expected to reach 0.03° at E > 1 TeV, thus boosting the chances of resolving some characteristic dark matter features. Furthermore, this extension would improve the projected (baseline) CTA sensitivity by a factor ∼3 in the 100 GeV–1 TeV energy range. With such configuration, and hundreds of hours of the Galactic Center observations, CTA would be able to exclude numerous dark matter models, significantly reducing the currently unexplored parameter space (Fig. 6.3).

Full Likelihood in the CTA Analysis Here are presented some predictions for the CTA sensitivity for signals from dark matter annihilation, when the analysis is performed using the full likelihood method. The considered scenarios include continuous spectrum of secondary gamma-rays, as well as the monochromatic line from direct annihilation into two photons (always assuming Br = 100 %). For the calculation of the response function, one of the CTA 2012 official MC productions is used (configuration E,[1] [8]).

Figure 6.4 shows the expected 95 % c.l. upper limits on $\langle\sigma_{\text{ann}} v\rangle$ from the Segue 1 observations with CTA, assuming annihilation into $b\bar{b}$, $\tau^+\tau^-$ and W^+W^- final

[1]The simulations correspond to the baseline Southern array (and do not include the CTA-US array enhancement), while Segue 1 is observable from the North. Still, the sensitivity curve scales linearly with $1/J$ for observations well outside the Galactic plane, and therefore are trivially computed for Southern targets of similar characteristics.

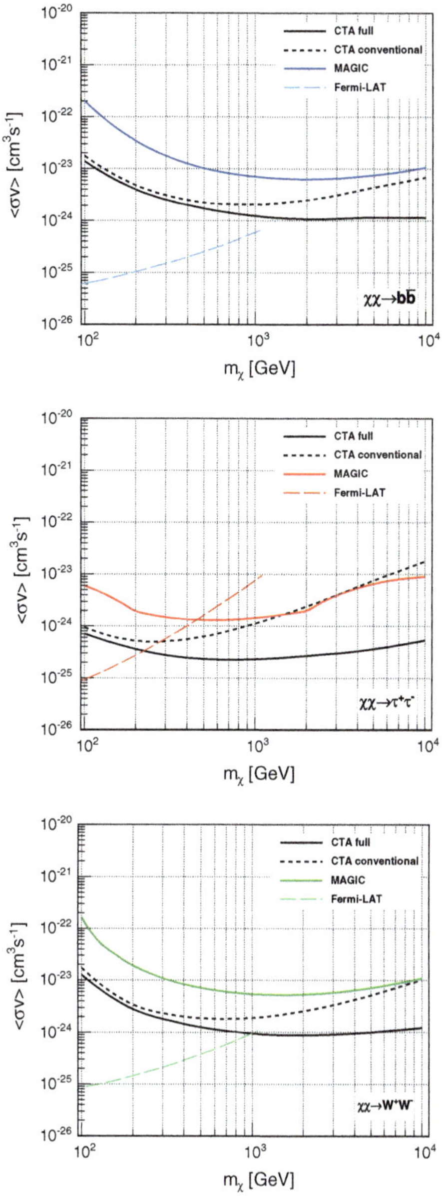

Fig. 6.4 95 % c.l. upper limits on $\langle\sigma_{\rm ann} v\rangle$ as a function of m_χ, for $b\bar{b}$ (*top*), $\tau^+\tau^-$ (*middle*) and W^+W^- channels (*bottom*), estimated for 150 h of Segue 1 observations with CTA, calculated with the full likelihood (*full black lines*) and with the conventional approach (*dashed black lines*). Also shown are the exclusion curves from MAGIC (obtained from this work Sect. 5.4.1.1, full *blue*, *red* and *green lines*, for the considered channels, respectively) and from Fermi-LAT (*long-dashed lines*, same color coding, [5])

states. For consistency with the MAGIC results presented in Chap. 5, these estimates are made for 150 h of Segue 1 observations. The number of background regions is taken to be $\tau = 12$, and the adopted value of the astrophysical factor J is 1.7×10^{19} GeV^2 cm^{-5} (Einasto profile, [7]). The most constraining bounds correspond to $\langle\sigma_{\text{ann}}v\rangle \sim 1 \times 10^{-24}$ cm^3 s^{-1} for $m_\chi \sim 2$ TeV, $\langle\sigma_{\text{ann}}v\rangle \sim 2 \times 10^{-25}$ cm^3 s^{-1} for $m_\chi \sim 700$ GeV and $\langle\sigma_{\text{ann}}v\rangle \sim 9 \times 10^{-25}$ cm^3 s^{-1} for $m_\chi \sim 2$ TeV, for the $b\bar{b}$, $\tau^+\tau^-$ and W^+W^- final states, respectively. Therefore, depending on the channel, for the considered setup and with the full likelihood analysis, CTA is a factor 7–30 away from the canonic $\langle\sigma_{\text{ann}}v\rangle$ value. These limits are about an order of magnitude more constraining than those from the Segue 1 observations with MAGIC (Sect. 5.4.1.1), which is as expected given the performance goals of CTA. Also shown is the comparison with Fermi-LAT exclusion curves from joint analysis of 10 dSph observations [5]: CTA bounds become more stringent for $m_\chi > 1$ TeV, for $b\bar{b}$ and W^+W^- channels, and for $m_\chi > 220$ GeV for annihilation into $\tau^+\tau^-$.

Figure 6.4 also shows the expected limits from the analysis with the conventional method. These are included for comparison, since this method is currently used as standard in the CTA analysis chain [7]. As it can be seen, the conventional approach is clearly suboptimal (for dark matter searches) compared to the full likelihood: the Improvement Factor values, for $m_\chi = 100$ GeV, 1 TeV and 10 TeV are 1.3, 1.7 and 6.0 ($b\bar{b}$ channel), 1.3, 4.8 and 33.1 ($\tau^+\tau^-$ channel) and 1.4, 2.0 and 8.5 (W^+W^- channel), respectively. Some of these Improvement Factors reach such high values partially because no energy range optimization has been applied for the conventional method. Nevertheless, even when the optimal range is used, the gain from the full likelihood is up to 70 % (see also Sect. 4.3 for predicted improvements assuming various BM models).

Prospects for line searches with CTA are estimated for observations of Segue 1 and of the Galactic halo (with $J = 3.3 \times 10^{21}$ GeV^2 cm^{-5} and $\tau = 2$ for the

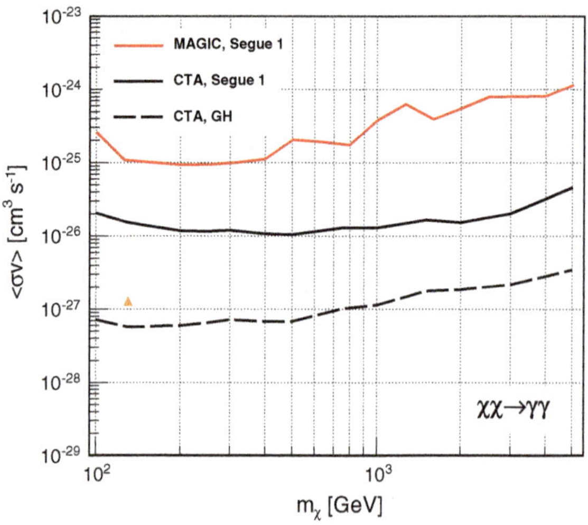

Fig. 6.5 Estimates of the CTA 5σ exclusion limits on the $\langle\sigma_{\text{ann}}v\rangle$ as a function of m_χ (*black lines*), with the full likelihood analysis of 150 h of Segue 1 (*full line*) and the Galactic Halo observations (*dashed line*). Also plotted is the MAGIC exclusion curve from this work scaled to 5σ significance (*red line*, Sect. 5.4.2.1), and the $\langle\sigma_{\text{ann}}v\rangle$ value corresponding to the 130 GeV gamma-ray line (*orange triangle*, [6])

latter, [9]). The considered dark matter particle mass range is between 100 GeV and 5 TeV. Figure 6.5 compares the CTA expected 5σ exclusion curves on $\langle\sigma_{\rm ann}v\rangle_{\gamma\gamma}$ from Segue 1 observations with that of MAGIC (scaled to 5σ from the 95 % c.l. upper limit presented in Sect. 5.4.2.1): the former is, depending on the energy, a factor 3–20 better than the latter. The strongest bound from CTA corresponds to $\langle\sigma_{\rm ann}v\rangle = 1 \times 10^{-26}$ cm^3 s^{-1}, for $m_\chi = 500$ GeV. The triangle marks the estimated $\langle\sigma_{\rm ann}v\rangle$ for which the hint of a line at 130 GeV is explained through $\chi\chi \rightarrow \gamma\gamma$ channel (Sect. 5.4.2.1, [6]); the corresponding CTA constraint from Segue 1 is about a factor of 20 away from probing this value.

On the other hand, with the observations of the Galactic Halo, the 130 GeV line claim can be tested with CTA in under 150 h (Fig. 6.4). The sensitivity in this case is a factor $\sim$30 better than expectations for CTA from Segue 1. However, the systematic uncertainties are not taken into the account, which might be relevant for this search, given how the gamma-rate would be $\sim$2 % of that of the background at the low energy end of the spectrum.

In any case, it should be emphasized that predictions presented in this chapter must be taken with some reserve. Further studies and design improvements are being implemented to the final CTA configuration, which could eventually change, in a significant way, the performance assumed here. Nonetheless, these estimates clearly illustrate the tremendous contribution that CTA could make on indirect dark matter searches in the upcoming years. In addition, the implementation of the full likelihood method in the CTA analysis can only increase its impact.

References

1. The CTA Consortium: The Cherenkov Telescope Array home page (2013), https://www.cta-observatory.org
2. M. Actis et al., Design concepts for the Cherenkov Telescope Array CTA: an advanced facility for ground-based high-energy gamma-ray astronomy. Exp. Astron. **32**, 193–316 (2011), arXiv:1008.3703
3. J. Sitarek et al., Physics performance of the upgraded MAGIC Telescopes obtained with Crab Nebula data. To appear in the *Proceedings of Proceedings of the 33rd International Cosmic Ray Conference*, Rio de Janeiro, Brasil (2013)
4. M. Wood et al., Prospects for indirect detection of dark matter with CTA. White paper contribution for Snowmass 2013 in the Cosmic Frontier Working Group CF2: WIMP Dark Matter Indirect Detection (2013), arXiv:1305.0302
5. M. Ackermann et al., Constraining dark matter models from a combined analysis of Milky Way satellites with the Fermi Large Area Telescope. Phys. Rev. Lett. **107**, 241302 (2011), arXiv:1108.3546
6. C. Weniger, A tentative gamma-ray line from dark matter annihilation at the Fermi Large Area Telescope. JCAP **08**, 007 (2012), arXiv:1204.2797
7. M. Doro et al., Dark matter and fundamental physics with the Cherenkov Telescope Array. Astropart. Phys. **43**, 189–214 (2013), arXiv:1208.5356
8. K. Bernlöhr et al., Monte Carlo design studies for the Cherenkov Telescope Array, Astropart. Phys. **43**, 171–188 (2013), arXiv:1210.3503
9. J. Aleksić, J. Rico, M. Martinez, Optimized analysis method for indirect dark matter searches with imaging air Cherenkov telescopes, JCAP **10**, 032 (2012), arXiv:1209.5589

Conclusion

The presented work reports on indirect dark matter searches in the dwarf spheroidal galaxy Segue 1 with the MAGIC Telescopes. Observations, carried out between January 2011 and February 2013, have resulted in 157.9 h of good-quality data, thus making this the deepest survey of any dSph by any IACT so far. In addition, this is one of the longest observational campaigns ever, with MAGIC or any other IACT, on a single, non-variable object. That imposes some important technical challenges, for which suitable and optimized solutions have been successfully designed and implemented.

Data analysis was performed by means of the *full likelihood method*, a dedicated approach developed as a part of this work. Full likelihood is optimized for recognition of spectral features, like the ones expected from dark matter annihilation or decay: e.g. monochromatic line, spectral cutoff or peak from the virtual internal bremsstrahlung photons. Furthermore, it was demonstrated that this method is more sensitive than the conventional analysis approach (currently standard for the IACTs), even for the featureless spectra: depending on the spectral slope, an improvement of up to 70 % can be achieved. Additionally, it was demonstrated that the full likelihood is unbiased, robust and stable, and that it allows a straightforward combination of results from different targets and observatories.

No significant gamma-ray excess was found above the background in the Segue 1 sample. Consequently, the observations were used to set constraints on the dark matter particle properties, assuming various final state scenarios. In particular, limits were computed for the spectral shapes expected for secondary gamma rays from dark matter annihilation and decay into Standard Model pairs ($b\bar{b}$, $t\bar{t}$, $\mu^+\mu^-$, $\tau^+\tau^-$, W^+W^- and ZZ), for monochromatic gamma-ray lines, for photons produced by the virtual internal bremsstrahlung and for the spectral features from annihilation to gamma-rays via intermediate scalars. The calculations were done in a model-independent way, by assuming a branching ratio of 100 % to each of the considered final states. 95 % confidence level limits were obtained for m_χ in the 100 GeV–10 TeV range and 200 GeV–20 TeV, for annihilation and decay scenarios, respectively.

J. Aleksić, *Optimized Dark Matter Searches in Deep Observations of Segue 1 with MAGIC*, Springer Theses, DOI 10.1007/978-3-319-23123-5

Table 1 Summary of the strongest limits and corresponding m_χ, obtained from Segue 1 observations with MAGIC, for various final states from dark matter annihilation (ANN) and decay (DEC)

Secondary photons				
ANN	Final state	m_χ (TeV)	$\langle\sigma_{\rm ann} v\rangle$ (cm^3 s^{-1})	Most constraining limit from dSphs
	$b\bar{b}$	2	6.2×10^{-24}	$m_\chi > 1$ TeV
	$t\bar{t}$	2.5	8.7×10^{-24}	$m_\chi > 1$ TeV
	$\mu^+\mu^-$	0.4	4.4×10^{-24}	$m_\chi > 0.3$ TeV
	$\tau^+\tau^-$	0.55	1.2×10^{-24}	$m_\chi > 0.45$ TeV
	W^+W^-	1.5	5.1×10^{-24}	$m_\chi > 1$ TeV
	ZZ	1.7	5.6×10^{-24}	$m_\chi > 1$ TeV
DEC	Final state	m_χ (TeV)	τ_χ (s)	Most constraining from IACTs
	$b\bar{b}$	20	2.6×10^{25}	$m_\chi < 0.9$ TeV
	$t\bar{t}$	20	2.0×10^{25}	$m_\chi < 0.9$ TeV
	$\mu^+\mu^-$	20	6.2×10^{24}	$m_\chi < 1.3$ TeV
	$\tau^+\tau^-$	20	3.1×10^{25}	$m_\chi < 0.65$ TeV
	W^+W^-	20	2.6×10^{25}	$m_\chi < 0.75$ TeV
	ZZ	20	2.7×10^{25}	No information
Monochromatic line				
ANN	Final state	m_χ (TeV)	$\langle\sigma_{\rm ann} v\rangle$ (cm^3 s^{-1})	Most constraining limit from dSphs
	$\gamma\gamma$	0.2	3.7×10^{-26}	$m_\chi > 0.13$ TeV
	$Z\gamma$	0.2	7.5×10^{-26}	$m_\chi > 0.2$ TeV
DEC	Final state	m_χ (TeV)	τ_χ (s)	Most constraining limit from dSphs
	$\gamma\nu$	12.5	1.8×10^{26}	$m_\chi > 0.2$ GeV
Virtual internal bremsstrahlung				
ANN	Final state	μ	m_χ (TeV)	$\langle\sigma_{\rm ann} v\rangle$ (cm^3 s^{-1})
	$\mu^+\mu^-(\gamma)$	1.05	0.25	8.8×10^{-26}
		1.50	0.25	1.1×10^{-25}
		2.00	0.25	1.2×10^{-25}
	$\tau^+\tau^-(\gamma)$	1.05	0.25	8.7×10^{-26}
		1.50	0.25	1.0×10^{-25}
		2.00	0.32	1.1×10^{-24}
Gamma-ray boxes				
ANN	Final state	m_ϕ/m_χ	m_χ (TeV)	$\langle\sigma_{\rm ann} v\rangle$ (cm^3 s^{-1})
	$\gamma\gamma\gamma\gamma$	0.10	0.25	5.5×10^{-26}
		0.50	0.3	5.8×10^{-26}
		0.99	0.4	5.6×10^{-26}

When applicable, it is stated for which m_χ these limits become the most constraining (either from dwarf spheroidal galaxy observations, or among the IACT results on any kind of source)

Table 1 shows a summary of the results obtained in this work for each of the explored processes and channels. When applicable, the results are also put in context with respect to those obtained by other instruments or from different targets. It must be stressed, however, that for all considered models, the limits calculated in this work are the most constraining from the observations of any dwarf spheroidal galaxy by any IACT so far.

Exclusion curves derived for secondary photons represent an improvement of an order of magnitude with respect to the previously most constraining MAGIC results. They are complementary to the Fermi-LAT observations and, for leptonic channels, even more constraining above $m_\chi \sim 400$ and 450 GeV, for $\mu^+\mu^-$ and $\tau^+\tau^-$ final states, respectively. Overall, depending on the channel, results from this work are a factor 40–300 away from the $\langle\sigma_{\text{ann}}v\rangle$ thermal value of $3 \times 10^{-26}\text{cm}^3\ \text{s}^{-1}$, if no additional boost factors are considered.

Monochromatic lines were also studied, assuming direct annihilation into two photons, or a photon and a Z boson, as well as the decay into a photon and neutrino. Although few orders of magnitude weaker than the currently most constraining bounds from the observations of the central Galactic region by Fermi-LAT and H.E.S.S., limits from this work are the strongest from dwarf observations and more stringent than the Fermi-LAT ones from observations of the same kind of objects, for $m_\chi > 130$ GeV.

Annihilation into charged particles accompanied by emission of a virtual internal bremsstrahlung photon was considered for leptonic final states and the mass-splitting parameter values for which the virtual photons peak is the most pronounced ($1.05 \leq \mu \leq 2.00$). It has been shown that the obtained bounds are very similar for both the $\mu^+\mu^-(\gamma)$ and $\tau^+\tau^-(\gamma)$ channels, as well as that the most constraining upper limit is rather independent on the value of the μ parameter, and is of order $\langle\sigma_{\text{ann}}v\rangle \sim 9 \times 10^{-26}\text{cm}^3\ \text{s}^{-1}$ for $m_\chi \sim 250$ GeV. Compared to the exclusion curve obtained without the virtual internal bremsstrahlung contribution, these limits are, depending on m_χ, a factor $\sim$ few to over one order of magnitude more constraining.

Other features that increase the detectability prospects are gamma-ray boxes—characteristic signatures whose spectral shape is completely defined by the masses of the dark matter particle and the intermediate scalar ϕ. In this work, results are presented assuming m_ϕ/m_χ values that lead to various box widths, from very wide to very narrow. The $\langle\sigma_{\text{ann}}v\rangle$ constraints achieved for these extreme cases were found to be in the (5–6) $\times 10^{-25}\text{cm}^3\ \text{s}^{-1}$ range, thus proving that box-like features can be quite relevant for constraining the nature of dark matter particle.

Altogether, results from this work are contributing to the field by improving the previous limits from dwarf spheroidal galaxies and by complementing the existing bounds from other targets. They also fulfill the original goal of this project, which was to improve the MAGIC sensitivity for dark matter searches by an order of magnitude, with respect to the previous most stringent results. This has been accomplished through the use of the stereo system, longer exposure and analysis optimized for dark matter signals.

The obtained exclusion limits are typically a factor $\sim$100 less constraining than predictions from the most natural models like, e.g. the constrained Minimal Supersymmetric Standard Model. Because of this, it was chosen to present the results in a model-independent way, by focusing on general spectral features rather that specific theoretical dark matter realizations. It should be kept in mind, however, that these results are somewhat conservative: no flux enhancements, due to possible boost factors, have been considered. In general, the astrophysical uncertainties entering the expected fluxes are large enough so that potential surprises cannot be excluded.

The complementarity of indirect gamma-ray searches with other techniques is the key for a thorough understanding of the dark matter nature. A positive signal from direct detection would poorly constrain the dark matter particle mass, whereas unambiguous recognition of a dark matter signal in the accelerators is a rather difficult task. Indirect searches provide unique ways for measurement of the dark matter spatial distribution in the Universe and its other most important parameters.

Studying different targets is of particular importance for indirect dark matter searches. On one hand, a certain confirmation of the dark matter signal, especially if it is a featureless one, can only come from observations of at least two sources. On the other hand, diversity among observational targets is necessary, as searches in different objects are affected by different uncertainties. For instance, although most aspects of the general cold dark matter halo structure are resolvable from numerical approaches, the current knowledge and predictive power regarding its behavior are limited by the complex interplay between the dark matter and baryonic components. It is still a long way until the full perception of the effects the baryons have on the dark matter distribution is achieved. This is particularly relevant for targets like the Galactic Center and halo, or galaxy clusters, since their significant luminous content can temper with the evolution of the dark matter component. Furthermore, there are also uncertainties coming from the presence of substructures in the halo, and the possible enhancements of the cross section due to the quantum effects, that directly influence the value of astrophysical factor J. These uncertainties are large ($\mathcal{O}(10)$ or more) and their impact on halos may be different on different scales. Therefore, diversification of the observational targets is the optimal strategy for the discovery.

The indirect dark matter searches with MAGIC will continue in the upcoming years. Deeper observations of Segue 1 or another suitable target will push the present sensitivity even further. At the moment, a proposal is being shaped to combine, through the full likelihood, the observations from different dark matter targets and gathered by the different IACT observatories: MAGIC, H.E.S.S. and VERITAS. Assuming equal exposures by each of these instruments, a combined analysis would lead to a $\sim$70 % more stringent results (compared to the results from each of the observatories individually), thus becoming the most sensitive, global effort in the indirect searches by the current generation of IACTs. Even more is expected from the CTA, whose enhanced possibilities combined with the optimized analysis—such as the full likelihood method—could bring us closer than ever to solving the mystery called dark matter.

Index

J. Aleksić, *Optimized Dark Matter Searches in Deep Observations of Segue 1 with MAGIC*, Springer Theses, DOI 10.1007/978-3-319-23123-5

CPSIA information can be obtained at www.ICGtesting.com
Printed in the USA
BVOW10*0322241115

428309BV00002B/6/P

9 783319 231228